The Keys to Our Success

Lessons Learned from 25 of our Best Project Managers

Compiled by
David Barrett and Derek Vigar

First Edition

Multi-Media Publications Inc.

Oshawa, Ontario

The Keys To Our Success:
Lessons Learned from 25 of our Best Project Managers
Compiled by David Barrett & Derek Vigar

Managing Editor:	Kevin Aguanno
Editor:	Susan Andres
Typesetting:	Carolyn Prior
Cover Design:	Francis Bomban
eBook Conversion:	Carolyn Prior

Published by:
Multi-Media Publications Inc.
1050 Simcoe St. North, Suite 110
Oshawa, ON, Canada, L1G 4W5
http://www.mmpubs.com/

Paperback	ISBN-13: 9781554891627
eBook	ISBN-13: 9781554891634

Published in Canada. Printed simultaneously in Canada, the United States of America, Australia, and the United Kingdom.

CIP data available from the publisher.

Table of Contents

Managing Teams

Business Outcomes

Change

Delivery

Stakeholders

Scope

Quality

Risk

Vision

The Keys to Our Success

List of Authors

Our Stories

Although we asked everyone the same question, we were amazed with the array of responses. It became clear that people who stand out in the profession attribute their success to a variety of keys. Whether the keys pertain to leadership, change management, client ownership, managing virtual teams, having fun, or twenty other unique skills, each author makes a compelling case for why his or her key is so critical to success. The benefit from all this variety is that we gain insight into what it takes to be a "complete" professional. It also allows us to say with confidence that there is something in here for everyone.

There is no shortage of project management literature out there, but we have worked to make this personal. Inside, you will find storytelling with substance. Short bites filled with benefits, real examples, and tangible takeaways. Although written specifically for project management, most of these lessons have greater application to business and even personal success.

Our "Keys to Success"

For us, the "key" became a useful and playful symbol to capture all we hoped to accomplish with this project.

- Whether it was the notion that you could turn the key and unlock secret insights into the profession, almost as if you were reading someone's personal journal.

- Or perhaps you had proved yourself ready to drive your projects, and this group was going over a few pointers before giving you your set of keys.

- Maybe you need to practice these keys in music repeatedly before you can show mastery and make your projects "sing."

The Keys to Our Success

Regardless the analogy, we hope that all project managers who read this book will learn, grow, and be better tomorrow for something they unlocked through these pages. These keys have helped enable twenty-five project managers be successful, and we hope they have the same result for you.

Sincerely,

David and Derek

David Barrett

Derek Vigar

Communication

The Keys to Our Success

CHAPTER 1

The Seven Bullets of Highly Effective Project Managers

By Benoit De Grâce

D oes the following scene sound familiar to you?

It's been a long day at work. You've been going from meeting to meeting, occasionally getting back to your office to get something done, only to be interrupted by phone calls and various queries from coworkers.

You scan the messages on your Smartphone and find one that seems important; however, it is long. You flick your finger repeatedly to scroll through four or five paragraphs, searching to identify the "so what." You decide to skip it and plan to read later when you have time to figure out what's asked of you.

Or how about this one?

• Your colleague has just sent you a list of requirements he wants you to validate to see if he forgot anything. "OK," you think, "no big deal. This should only take a few minutes." The document is a spreadsheet. You open it, and then it hits you in the face like a ton of bricks. The spreadsheet is 283 lines long and has 22 attribute columns. Wow! Where do you begin?

Information overload, we call this. It eats away at your day. It causes you to miss important details. It throws "speed bumps" on your road to project success. Personally, it took an embarrassing Christmas party mishap to force me to look into a solution to this problem.

As one of the two owners of my consulting firm, you see, I make sure that each year we hold a Christmas dinner and party for our staff and their spouses or significant others. OK! I don't organize this event; my partner and I have an assistant who does a much better job than we ever could. So, one busy day in early December, I get an e-mail from our head office administrator with details to the event. It describes the theme, the venue, the time to get there, and the agenda and sprinkles in a few jokes, fun facts, and so on.

Finally, the day of the party, I get in my car and drive 200 kilometers to get to the venue, only to discover everyone had brought a gift! And here I am, the senior vice president, showing up empty-handed.

"Did I miss something here?" I asked our assistant. "Did you not get the message?" she replied.

Typical! Buried at the end of the e-mail was a note that said, "We are having a gift exchange, so bring a funny gift of no more than $10 in value!" Problem was, of course, I never even scrolled down far enough to read it.

(I was fortunate enough, though, that even at such a late time, a shop across the street was still open where I could get two giftwrapped miniature bottles of ice wine for just $9.99! Saved the day and got a few laughs from the staff!)

That got me thinking. This sort of miscommunication happens all the time in our projects. Then, I remembered that early in my career as a consultant, I read in a book that when companies decided to standardize phone numbers in North America, they looked into how long a string of numbers most people could reliably memorize. It turned out to be seven. Hence, the seven-digit numbers we have today (OK, with a 3-digit area code, but forget this bit of information for a minute).

I started to call this the "Phone Number Rule." It states, "Because people can only reliably memorize or visualize sets of seven things at most, it is usually good practice always to group ideas and other different items in sets of about seven."

Out of curiosity, I started to look into this to see whether it could have other practical uses, especially for me as a project manager. Maybe this Phone Number Rule can help improve how we communicate in projects and save more than just a trip to the gift store.

Lucky Seven

"What about the magical number seven? What about the seven wonders of the world, the seven seas, the seven deadly sins, the seven daughters of Atlas in the Pleiades, the seven ages of man, the seven levels of hell, the seven primary colors, the seven notes of the musical scale, and the seven days of the week?"[1]

---

[1] Miller, George A. "The Magical Number Seven, Plus or Minus Two." *The Psychological Review.* 63 (1956): 81–97.

In his paper from 1956, Miller goes on to explain "stimuli," "variance," "bits," "recoding," and other technical jargon. In the end, however, he withholds judgment about why this number keeps popping up everywhere.

For what it's worth, and I must stress I have no scientific evidence to back this up, I believe it has to do with anatomy. I've come to think that it might have to do with the fact that we are born with ten digits and that our brains have become wired by evolution to recognize what "10" looks like in an instant. Or maybe it's something deeper, but I'll leave that to the neurobiologists.

There's plenty of evidence that this "magical seven" is powerful. Marketing experts are quite good at using it, especially editors and publishers. Why do you think practical and self-improvement books have titles such as *The Seven Habits of Highly Effective People* or *Seven Steps to Happiness?* (OK, so the title to this book, The Keys to Our Success: Lessons Learned from 25 of Our Best Project Managers, is the exception that confirms the rule... but we were asked to keep each of our chapters to around "seven pages.")

If you want to convince yourself that this is true, try this simple experiment with some friends or family members (they have to be old enough to count without their fingers!). I have tried something similar with pictures on a slideshow with amazing results.

- Drop four different coins on the table (a penny, a nickel, a dime, and a quarter).

- Put a cloth or piece of cardboard over the coins.

- Gather your friends around the table and tell them to pay close attention, as you are about to reveal something.

- Remove the cloth (or cardboard) for a second and put it back.

- Ask what they saw.

Now, tally the results. How many got the number of coins right? How many got the kinds of coins right?

Try it again, varying the number and types of coins each time, and notice when people start to "guess." My prediction is that people will generally remember seven to nine of the coins; any more than that, people start to guess, and results become unreliable. How can we apply this to our practice as project managers? Over the years, I have found four very concrete and tangible applications of this Chunking Principle for project management. Let me share them with you.

The Chunking Principle

This phenomenon is also known in instructional design circles as the Chunking Principle. So, years later, as I began to develop and deliver project management courses, I read about how it is best to design slides to have no more than seven bullets when working in PowerPoint. There's that number seven again!

Briefly, the Chunking Principle says you should avoid the following:

- More than seven bullets on a slide
- More than seven bullets on a bulleted list in any document
- More than seven bubbles on a bubble chart
- More than seven columns on a spreadsheet
- More than seven slices in a pie chart

(By the way, did you notice how my bulleted list in the coin exercise had only five bullets? And this one had only five?)

Work Breakdown Structures.

How does the number seven apply to the WBS, you ask?
Simply put: "You should design a WBS in such a way
never to have, in any of its branches, more than five to ten
deliverables or work packages."

So, the following WBS is fine:

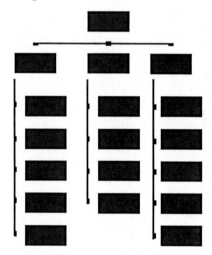

Whereas, this one violates the rule:

Notice how even though both have fourteen work packages (the lowest level boxes in each branch), it's much easier to see this with the first model than with the second. You have to count how many boxes there are in the second branch of the second WBS. That's the power of the Chunking Principle and it makes communicating your WBS much easier and more reliable.

This also proves useful when editing a WBS. People will more easily pick up if something is missing or duplicated in a diagram that complies with the Phone Number Rule.

Written Documents

The same rule applies to various documents. As project managers, we are told that we spend or should spend up to 80% of our time communicating. If so, we had better make sure people understand what we try to communicate. Otherwise, that's much room for errors and misunderstandings, which leads to huge productivity losses both at a personal and team level.

Writing paragraphs like this one is fine for a white paper, an article, or a book. But when dealing with technical documentation, keep in mind this other tip someone once pointed out to me: "People don't read; they scan."

That's right; we don't read—unless we expect the writer is trying to tell a story of some sort (the kind I am writing now). If we don't expect a story, we just scan for familiar patterns, key words, and ideas.

There are many reasons for this. They range from the fact that reading on a computer screen is tiring for the eyes and slower than reading paper (about 25% slower according to some) to our habit of surfing the Web to the pace of modern life.

Here is more evidence we don't systematically read everything. Ever received this in an e-mail: "Arocdnicg to rsceearch at Cmabrigde Uinervtisy, it deosn't mttaer in waht oredr the ltteers in a wrod are, the olny iprmoatnt tihng is taht the frist and lsat ltteer are in the rghit pcale. The rset can be a toatl mses and you can sitll raed it wouthit pobelrm. Tihs is buseace the huamn mnid deos not raed ervey lteter by istlef, but the wrod as a wlohe." Funny how your brain can still figure this out?

The same is true for a paragraph. Your brain reads a block of text and tries to recognize patterns and ideas. The longer the paragraph, the more chances your brain will get it wrong. Keep your writings to short lists of one- to two-line bullets.

Gantt Charts

Every now and again, I hear this complaint from people exposed to tools such as Microsoft Project or Primavera P6: "These tools are way too complex for us. I don't understand what all these figures mean! Could someone please design a simpler tool?"

After close inspection, though, I've noticed that most of these complaints come from the same type of users. Highly experienced planners who don't think twice about printing Gantt charts with sometimes up to fourteen or even twenty columns of data are exposing these people to these tools for the first time. Everything from Task ID to Early and Late Dates, Earned Value figures, CPI, Labor Cost, Total Float, Resources, and so on. Their thought, as experienced users, is "the more info on the same page, the less time I spend looking for it."

But the people getting these reports do not have brains wired for this kind of information. They are overloaded.

If you want to get your message across clearly, stick to no more than five or ten columns. This is definitely a case of less is more.

E-mails

Finally, let's get back to our old friend from the beginning—
e-mail—all those long e-mails we send one another, so
long that we don't take the time to read them! We scan the
paragraphs looking for the information we expect to see. We
skip important details because they are buried in a sea of
gibberish and unnecessary words. This is where I feel the
Phone Number Rule becomes so powerful.

Here is the Chunking Principle applied to e-mail for
efficient communication (aside from all the standard rules
about e-mails, such as "to whom to send the message,"
"avoid using reply all," and "avoid using capital letters," of
course).

- Give the message a subject line that states exactly
  what you expect as a response or the e-mail's
  purpose. Write your e-mail subject line as you would
  a newspaper headline.

- Start the e-mail with a general greeting. This
  paragraph is polite, but let's face it; it probably won't
  get much attention. Don't put anything important in
  it.

- Write the core of the e-mail, the important stuff,
  as a series of no more than five to ten bullets, each
  having no more than two or perhaps three lines.

- If you're having trouble writing your e-mail in this
  format, or if you're getting beyond five to ten bullets,
  pick up the phone!

That last bullet is a sign that what you are trying to
communicate is too complex for e-mail. In this case,
you should consider communicating in a face-to-face
conversation or, at the very least, a phone call. Remember
that even phone calls come complete with the instant
feedback and nonverbal cues that unfortunately do not
come with e-mail.

In a Nutshell

Remember that the human brain can only reliably understand information when chunked in groups of about seven items at a time. Remember this tip when:

- Developing a Work Breakdown Structure. Try to stick to no more than seven items in each branch at each level.

- Written documents. Keep your bulleted lists to no more than seven items of one or two lines each.

- Gantt charts. Avoid having spreadsheets with more than seven columns.

- E-mails. Use a bulleted format, and if you can't stick to this format, pick up the phone!

Undoubtedly, this tip has helped me avoid many unpleasant situations, and I'm sure you'll use it in all sorts of situations. Perhaps you'll even begin to convert your colleagues, and they, in turn, will stop sending you all those irritating and long e-mails!

Unfortunately, this won't do anything to help with spam or chain letters! Maybe in our next book...

Benoit De Grâce

Benoit is the president of PMC—Project Management Centre, a firm that helps its clients deliver better projects faster and at a lower cost by providing project and portfolio management consulting and training since 1996.

Benoit has a BEng in Industrial Engineering. He is a member of the OIQ and a certified Project Management Professional (PMP) member of the Project Management Institute (PMI). His work has taken him across Canada as well as the United States and overseas with clients such as Bombardier, Caterpillar, Sandoz, Shell, Standard Life, Ultima Foods, the Ministry of Transportation, and the Department of National Defence.

Over the past twenty years, he has gained an excellent reputation as a trainer in project planning and control techniques, leadership, communications, and various project management tools, including Microsoft Project and Primavera P6. As a senior business analyst with a keen ability to synthesize business processes and adapt project management expertise to various corporate environments, he has helped organizations define and streamline their business processes. He has delivered more than four thousand hours of training in project and portfolio management.

Benoit De Grâce, ing, PMP

President
PMC—Project Management Centre
2, Place du Commerce, bur 220
Île-des-Soeurs (Montréal), QC
Canada, H3E 1A1

E-mail: benoit.degrace@pmc.ca

Website: http://www.pmc.ca

LinkedIn: Benoit De Grace

Twitter: @BenoitDeGrace

Phone: 514-766-7459 x214
Phone: 514-984-8248
Fax: 514-766-4269

Leadership

The Keys to Our Success

CHAPTER 2

Leadership Is Taken, Not Given:
Establishing, Maintaining, and Regaining Control of Projects

By Gus Cicala

Though people have used this quotation to the point of cliché about everything from sports to dog whispering, I have come to see how it has specific relevance to our role as project managers. Given that the success and efficiency of a project ultimately reflects on our ability to drive the initiative, we obviously are in a leadership role. Yet, the matrix structure makes this standing tenuous, as there is very little tangibility to our control over the project.

We don't have traditional leverage, such as hiring and firing, and we're not even in direct interaction with the day-to-day delivery of the project in a way that allows us to "lead by example." We're not even a permanent presence in the organization! This makes leadership almost a paradox for project managers, so the saying, "Leadership is taken, not given" has a very literal, almost concrete, truth as a project manager.

Controlling a Project through Rhythm and Cadence

In 1992, during my time with IBM, I was involved in a project that was so infamous, it was still used in training more than a decade later as an example of how to deal with a failing project. It has come to be known simply as the Bridge Project. This project was doomed from the beginning with everything from a scope that only allowed 67% of the necessary budget to crucial scheduling problems. But learning comes through experience, and the most valuable lessons come through negative experiences, so it should come as no surprise that I learned much about how to be the leader of a project through the Bridge Project.

With the project clearly out of my control, our most senior local executive offered me the "privilege" of regularly meeting with him one on one to discuss how the project was going. In these meetings, this executive taught me about how projects are like tennis matches—whatever balls are on our side of the net are our responsibility to return to the other side; whatever is on the other side is out of our hands. So, going into the next two or three meetings, I simply made sure that all balls were kept off my side of the net.

The specs of the Bridge Project were so generic that even when we satisfied them, there was still an end-of-project mess, full of open issues to be resolved. Before I took leadership of the project, the client dominated the meeting by listing all issues that weren't to their satisfaction. We developed a list of open action items that collated all the disparate loose ends into one master reference sheet. We were clear with the client that whatever was on that comprehensive list of items would be completed; whatever wasn't on it would not be completed. We outlined next steps for each action item, and we made sure that we did all the next steps on our end, assuring that all balls remained in their court.

An amazing thing started to happen in these meetings. Whereas the client previously dominated the meetings with their complaints, we started to dominate the meetings with our questions. Projects are much like negotiations because they're so nebulous up front, but then you flesh things out and find a middle ground as you go. And there's a saying in negotiations that whoever asks the most questions wins. We took leadership of the meeting by listing all balls left on their side of the net that were stalling the project's success. There were several examples of this. We were in charge of an IBM machine that printed tickets for a ferry. The client would say that the customers were dissatisfied with how quickly the tickets were printing. We reprogrammed the computers on the spec that they needed to be printed within five seconds. When the client came back and said they still weren't printing fast enough, we said that they were printing to the spec.

We asked for a show of hands of who had test cases showing that it wasn't printing at satisfactory speed, and there was nothing but a room of full pockets. We'd make requested changes and ask them for feedback. We'd ask about reviewing documents still pending their feedback. They'd complain about open issues, so we'd ask about next steps, telling them we couldn't move forward until they told us what to do. In short, we kept the balls in their court, and it was clear to them that we couldn't move forward until they were returned to us.

When you aren't taking care of your own issues in an organized fashion, then all you are doing is pushing blame back and forth, which only makes both sides imperceptive to what they could be doing themselves to make the project run smoothly. When you leave no room to be questioned, you can ask the questions, and you'll find that you'll get results. I've found in my twenty years of project management since the Bridge Project that this is the most pain-free way of taking the reins of a project.

Control: An Objective Measure

When you ask a project manager whether he or she has control over a given project, they will generally respond subjectively. They'll say, for example, "I think it's going well." Note, however, that, in the example of the tennis game, control is entirely objective. When we have all the balls off our side of the net, and they are returned regularly, we control the project. More specifically, when the execution processes are on a regular, cyclical basis; when there is a weekly (or monthly) cycle for collecting actuals of cost, labor, and schedule; and when we're analyzing variances to the baseline, we control a project. When we are not accomplishing these things, we are not properly leading the project toward success.

One reason people confuse project control as a subjective measure is because some degree of luck goes into every project's success, independent of how much control the project manager has. In other words, being in control of a project does not guarantee perfect execution, and being out of control does not guarantee failure. In this way, project management is much like parenting. We all know someone who turned out OK despite being raised poorly by his or her parents, and we all know someone who faced challenges in their life despite a good upbringing. This does not mean that you should try to get away with being bad parents, and it does not mean that you should test your fate by letting projects get out of control! Tracking actuals does not necessarily preclude the variances from becoming negative, but as long as you know when the variances are at risk of becoming unfavorable and attempt to determine and rectify the causes, you are giving the project the best chance at success. Likewise, not having any idea of what's going on with the project can sometimes turn out OK, but don't confuse good results with effective leadership. Out-of-control projects that succeed do so *in spite* of their project manager, not because of them.

So even though it is far more nuanced to measure "how" a project is going—and even though this measure is not even necessarily a reflection on the project manager—control over a project is cut-and-dried. Fulfill the objective measures of controlling a project, and you give it the best chance at success.

Maintaining Control through Front-Ending the Pain

Keeping the project in rhythm and under control minimizes the need for any confrontation, and ideally, we will not need to intervene at all. However, the ideal scenario is seldom the case, and problems inevitably arise. Being a leader means being the first to sight these unfavorable signs, and relaying this information will make us unpopular among our clients.

This is one of the most difficult aspects of managing a project. This aspect of our job is counter-instinctive. While it is human nature to seek pleasure and avoid pain, it is the imperative of risk management to seek pain actively. To take it one step further, not only should we actively seek pain, but we should also seek areas of potential pain, acting on problems that aren't even problems yet. We avulse pain from its more comfortable status as a far-off pain that might not come to fruition and force ourselves to experience it in the present. By doing so, we have to play the role of messenger to those likely to be in denial or, worse yet, to blame us for the news.

This will feel like the wrong thing to do to all but the most experienced project managers because our intuition will naturally try to persuade us to leave this discomfort for the future. However, it is crucial to front-end this pain as much as possible to maintain the project's rhythm. When we are proactive, we can revise the plan, re-baseline, and so on; when we're reactive, budgets are exceeded, deadlines are missed, and specs aren't met.

In other words, the earlier you identify potential concerns, the more options you'll have for assuring the project doesn't fall out of cadence. If you procrastinate the pain, then instead of wondering what might go wrong, you'll wonder what did go wrong.

Conclusion

Though it is common knowledge that leadership is a central pillar of project management, it is less easily expressed how to lead as an outside entity to the organization. As the saying goes, you need to "take" the leadership because it is not given to you organically through the structure of the company. The least controversial way to take leadership is to establish control of the project simply through keeping it in rhythm. When you act on your duties with regularity, those you are working with will naturally tend toward following that rhythm. When there's risk of falling out of rhythm, you need to confront the potential "pain" preemptively and explicitly to have the best chance of delivering the project on time, on budget, and with high quality.

Gus Cicala

Gus is the president, CEO, and founder of Project Assistants, Inc. He is a Wharton Business School graduate with more than twenty-five years of project management experience and thirty-years' worth of information technology development and consulting expertise. During his time with Project Assistants, he has brought his project and portfolio management expertise to an array of clients across the US and Western Europe. Gus is a popular speaker and writer on project and portfolio management and Microsoft Project as an enterprise project-management technology platform. He has published many popular articles and books on the subject, including contributions to Macmillan's Que book series: Special Edition: Using Microsoft Project. He is also a contributing author to the third edition of Expediting Drug and Biologics Development and a member of Microsoft's product advisory council.

Gus Cicala

1521 Concord Pike, Suite 301
Wilmington, DE
USA, 19803

E-mail: guscicala@projecassistants.com

Website: http://www.projectassistants.com

Phone: 302-477-9711

The Keys to Our Success

CHAPTER 3

No Virtue Shall Go Unpunished

By Chip Clark

I once had a colleague who kept a running list of what he called "The 10 Laws of Life." During the time we worked together, "No Virtue Shall Go Unpunished" sat at the #3 spot on the list, and his usual explanation involved a shake of the head, a lopsided grin, and the words "Life, my friend, is an apple-pie-eating contest for which first prize is... an apple pie." This is especially relevant to project managers, whose reward for completing a difficult, resource-constrained, multistakeholder, schedule-slipping nightmare of a project is often the assignment of another—even more daunting—project. When things go well, people tend to talk about the project being successful; when things don't go so well, they often attribute failure to the project manager and/or project team. Project management is definitely not for the faint of heart.

I have worked with dozens of project managers over the years, and the good ones have all mastered the tools of the trade. However, the **great** ones have done something else that is the key to both their success and their enthusiasm for their careers. They have learned to make project management fun—not the carefree, playful fun of childhood, but the grown-up business fun that comes from succeeding against all odds. They have mastered "Law of Life" #3 by finding the fun in as many aspects of project management as possible. They have learned, in effect, to love apple pie whether it is part of the contest or the prize.

How have these people developed a capacity for making project management fun? Three things stand out:

- They know and use the adage "different strokes for different folks."

- They take project risk assessment a step further, create personal "most likely" scenarios, and compare the results with their predictions.

- They focus—actively—on the personal side of professional development.

These keys to success are rooted in the mindset and personality of the project manager. It's time for some reflection.

Know Thyself

Early in my career as a consultant, I worked extensively with project managers in California's Silicon Valley. Most of their projects pushed the envelope of existing technology, creating leading-edge products ("bleeding-edge," they called it affectionately). On one memorable occasion, I facilitated a project closure/debrief session for a product development team at a well-known company in the computer industry. They had recently celebrated—with great fanfare—the launch of a new product.

We began the day by creating an "oral history" timeline of major project events. I asked them to put away all documentation and to work from memory, assuring them we would compare their recollections to the project history later in the day. Two team members drew the timeline and took notes on a wall-length whiteboard while I asked questions and attempted to keep the conversation organized.

We started the timeline when the "mandate" had emerged from an executive team meeting, and we ended it the day the product was "released" to the sales and marketing groups. As we worked through their recall of an eighteen-month development effort, a clear pattern emerged. The project manager consistently noted interim points of closure—decisions made, milestones met, problems solved, deliverables checked off. The software team leader consistently focused on information gained—valuable things learned from mistakes or false starts, insights that emerged from contentious cross-functional meetings, unexpected breakthroughs, and critical epiphanies about customer needs.

Over lunch, we talked about their different—and highly complementary—perspectives. They had figured out early that different aspects of the project motivated them and that this could work to their advantage. The project manager lives for the satisfaction of closure, tangible accomplishments, and checking things off his lists (both actual and mental). New information that initially seems to slow progress annoys him. The software manager is a self-described "information junkie" who seems genuinely delighted by what he learns at each stage of a project. Meeting deadlines—he said it almost apologetically—brings him very little satisfaction.

Together, they made a powerful team. The project manager's love of closure kept the project on schedule, and the software manager's love of learning helped the team integrate critical new information along the way. Perhaps more important, each was passionate about project management, loved his job, and looked forward to the next "mission impossible" assignment.

This is a classic example of finding the fun in project management through self-knowledge. Obviously, project success requires both meeting deadlines **and** integrating new information. And over the life of the project, each of these people spent time doing things that were not personally motivating. We all do. But they learned to make the most of the **fun** aspects of a project and to respect and trust their differences. They turned "different strokes for different folks" into a project advantage.

Key to Success: What do you love most about project work? What aspects do you find truly motivating? What aspects—necessary though they might be—seem tedious or outright unpleasant? Think about these questions as you plan the next project, and keep an eye out for people motivated by the very things that turn you off.

Project Prophecy

"Call anytime," I had said. But I wasn't expecting to hear from him two days after Christmas, phoning from a resort in Mexico where he was vacationing with his family. We had talked a week earlier, sending documents back and forth, as he finalized plans for a major change-management project that involved closing a facility in one city and transferring programs and some employees to a new building in another city. All this was scheduled for mid-January.

Everything seemed in place. He was an experienced project manager, the CEO was the project sponsor, the business case was clear, and we had gone over the plans, schedules, communication protocols, and risk analysis with a fine-toothed comb. More than once. He was confident about budgets and people resources.

But something that he couldn't quite articulate was bugging him, and he needed a sounding board. I asked him to think of the project, not as the sequence of tasks, activities, and events we had carefully planned, but as a narrative, a story with a plot. How did he think the narrative would unfold? Where was the tension? Would there be conflict? Heroes and villains? Resolution?

He was quiet for a minute and then began the project story: "The first three weeks will be fine. Closing the facility will mostly happen on schedule. The actual move will be exhausting, but we're as well prepared as we can be."

He went on through the schedule creating a detailed scenario, predicting how things would go. As he talked, and I asked more questions, he identified a stakeholder whose input he didn't trust and a simmering political issue involving the general manager whose division was being moved. We explored worst case/best case/most likely case scenarios for how these potential stakeholder issues might affect the project.

As things played out on this project, his instincts were 50 percent accurate. One of the two stakeholders in question tried to sabotage the project at the last minute. Because we had identified this possibility, the project manager saw it coming, responded immediately, and got the project back on track. And he **loved** that one of his predictions was correct. Thinking of a project as a story and predicting the narrative arc became regular parts of his preparation. He found fun both in generating predictions and in being right when some of them came true. His accuracy improved over time, and his ability to head things off at the pass contributed to a record of accomplishment of consistent project success.

Key to Success: Add a personal dimension to project risk assessment. Beyond the cost and schedule risks, the stakeholder uncertainties, the balance of project requirements and available resources—what does your gut tell you? What do you think will happen? Write this in a personal project journal or talk it through with a trusted colleague. How will the story of this project unfold? Add your knowledge of the human side—think in terms of conflict (and resolution), heroes and villains, narrative tension, and happy endings. Compare your story to actual project results and celebrate your ability to predict accurately; learn from each project and its story. Become a reliable project prophet.

Protect Your Sanity (and Your Career)

The first time someone with twenty years' experience managing large construction engineering projects showed up in an introductory Fundamentals of Project Management seminar, I was surprised. I probably blurted, "What are you doing here?" But it has happened so often now that I've almost come to expect it. One quality the best project managers share is a natural pursuit of continuous improvement—they never stop learning.

They treat every project, however small, as a chance to hone their skills. They sign up for classes and seminars because "You never know—you might learn something!" They take on home improvement projects and create a set of requirements and success criteria for family vacations. They find fun in treating every project as a personal learning/professional development opportunity.

I asked one of these "veterans" how he had developed the ability to so obviously enjoy his work. He told me this story.

Some twenty years earlier, as a newly graduated engineer, he took a job with a subsidiary of a large aerospace company. Almost immediately, he was assigned to a project team. The project manager (regarded as the best in the firm) developed a healthy cross-functional team.

It was an interesting project with some complicated technical challenges requiring input from all sides. Team meetings were lively, the project manager encouraged respectful disagreements, and project decisions were timely and clear. The team, in his words, "worked hard and played hard." Everyone pulled together to achieve project goals, and they were making very good progress.

Meanwhile, beyond the team, the world of corporate politics went on, and the subsidiary president lost a battle. In the frenzy of cost cutting and restructuring that followed, the project was canceled—abruptly. The team members were shocked, devastated, and angry, and they quickly became bitter. Their good work had all been wasted—stupidly, it seemed. Why bother investing personal energy in a project that can be canceled by people who know nothing about it? "Those idiots don't have a clue," they told one another. They were quickly working themselves into a very negative attitude.

The project manager scheduled a final all-day team meeting and asked a human resources manager to facilitate. First, they vented, and then they grieved. Finally, they conducted a routine project debrief—summarizing accomplishments, analyzing delays and missteps, identifying lessons learned. The project manager pushed each of them to focus on hard-won information. He challenged them to identify how they would use this experience to contribute to their next project team. He asked each of them to imagine stepping into the role of project manager—what would they do differently if they were standing in his shoes?

They had a conversation about engaging in the work for its sake, considering a positive outcome as a bonus, not a given. He was adamant that their work had not been wasted; it would add value to every project they worked on for the rest of their careers. "Use the lessons," he urged them. "Information is always valuable. **Use the lessons.**"

At least one member of that team truly took this advice to heart. Twenty years and dozens of projects later—at the pinnacle of a successful career—he enrolled in a fundamentals seminar. He came for a refresher, expecting to learn something, and because he thought it would be fun to meet some "budding project managers." "It's all in the learning," he told the other participants in the seminar. "Focus on the learning. You can't always rescue a doomed project, but you can rescue your sanity because you learned something useful."

Key to Success: When the plans and schedule are in place for your next project—before work on the first milestone begins—ask yourself these questions:

- **On this project, what is the most important thing for you to focus on?**

- **What is this project most likely to teach you?**

- **What aspect of this project is likely to bring you the greatest satisfaction?**

- **What aspect are you most likely to find frustrating?**

- **How will you personally celebrate when this project is completed?**

Record your answers. At project closure, compare your answers and predictions to your project experience. This is a win/win activity. If your predictions were accurate, celebrate that. If they weren't accurate, you've learned something new about yourself or project management or both. Celebrate that as well.

As a project manager, you will do things that play to your strengths; you will also do things that don't; some things will be energizing and fun, others tedious and exhausting. Notice the patterns of what motivates you. Focus on enjoying the energizing and fun parts. Use your experience to predict project outcomes. Learn to appreciate all the lessons a project can teach you. Remember that professional development is also personal development. Enjoy your career. And the next time someone tells you no virtue shall go unpunished, smile knowingly.

Chip Clark

Chip is an organization development consultant who first learned project management in 1989, working with high-tech project managers in California's Silicon Valley. She spent the next several years teaching project management skills to product development teams at companies from A (Apple Computer) to Z (Zonare Medical Systems). Project management training and consulting has been integral to her practice since.

She currently teaches Fundamentals of Project Management for the Gardiner Center at Memorial University in St. John's and the Interpersonal and Team Communications module for Master's Certificate in Project Management at the Schulich School of Business.

Chip's consulting experience includes facilitating project launch activities, coaching individual project managers, diagnosing and resolving critical project issues, and conducting in-depth project debriefing sessions. She is especially deft at bringing project management tools and discipline to strategic planning and change management efforts.

Chip is constitutionally incapable of working without humor, stories, practical approaches, and realistic solutions. She holds degrees in biology and counseling psychology, and she pursued graduate studies (and fieldwork) in cultural anthropology, which might seem far afield from project management, but it provides a surprisingly useful background.

Chip Clark

Pathfinder Management Group
179 Gower St.
St. John's, Newfoundland
Canada, A1C 1R1

E-mail: chip.clark@pathfindergroup.ca

LinkedIn: Chip Clark

Twitter: @gowermaven

Phone: 709-722-2127

The Keys to Our Success

CHAPTER 4

Know Your Leadership Style

By Catherine Daw

I am a sailor; I like the challenges of moving a sailboat through the wind and water to reach a destination. The thrill is in the journey and who is along for the ride. It takes teamwork and well-honed leadership skills. When you are the skipper, everyone must take his or her direction from you. It doesn't mean he or she can't ask questions or make suggestions, but at the end of the day, the skipper decides or guides the boat and crew through a safe journey to the destination.

Every year, we take a sailing trip somewhere around the world. In the end, the best sailing is in my backyard—Georgian Bay, Ontario, Canada. Each time I step on the boat, there are new lessons to learn, including many related to my leadership style, how to maximize my abilities as well as those of the crew, adapt my skills and style, and apply them to the particular tasks.

The lessons of leadership at the helm of a boat can easily be applied as a project manager—when are you an autocrat (pull that main halyard now!), participative (what do you think about the coming storm?), or when do you delegate (you go figure out how to fix the tangled spinnaker—I trust you to get it done)?

As project manager early in my career, I had luxuries that rarely exist today. My team was usually dedicated and focused only on the project (often seconded from various parts of the organization). Typically, we were co-located in one area or at least in the same building. Project complexity was much lower—either all technical or all business. The focus was much more on managing the project and overseeing the work done by a homogeneous and focused team.

With increasing complexity; a virtual, overallocated team; pressing needs from global competition, sophisticated organizations, clients; and economic pressures to deliver effectively, the luxury of focusing only on the technical and management aspects of a project are long gone. Figure 4.1 shows that, as you advance in your career and the complexity of projects increases, the need for leadership abilities climbs exponentially.

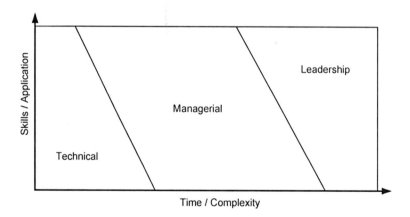

Figure 4.1: How leadership skills become more critical as project complexity increases.

Anyone adept at successfully overseeing projects realizes the way to get things done is through people. More than 80 percent of your role as a project manager is about people. Over my career, I have realized how important leadership style is in driving you toward success or failure. Even more, with increasing complexity, the need to apply exceptional leadership skills is more urgent, making it essential to know our style and figure out how to apply and adapt to situations and the needs of a project and its players.

Over time, I have learned my style changed, depending on needs—what to dial up or turn down. On a large program where I came in as a turnaround project manager, the team had struggled because of confused leadership, lack of direction, and minimal appreciation of their efforts. Using my leadership strengths and leveraging my style to consider the situation, I rallied the team and turned them from a dysfunctional group into a high functioning team, turning the project from red to green faster than other parts of the program. We became a beacon of light in a stormy program environment. Our project advanced despite a reframing of the overall program, which led me to believe firmly in the power of leadership and its practical application in any project or program situation.

How to Know your Style

Theories abound. I have learned to keep it simple. How do we best determine our style? Most of us have taken some form of assessment that considers our leadership style. Leadership 360 profiles and similar tools give you a frame of what your leadership style might be.

Sticking to the basics and considering many simple elements is at the heart of knowing more about your style. Here are four different ways I have found help better understand our style and use those frames to make the style adjust in the wide range of situations you will find yourself.

Three core styles

1. Authoritarian or autocratic

2. Participative or democratic

3. Delegating or free rein

I have learned that good leaders use all three styles with one typically dominant. Generally, bad leaders tend to stick with one style. Let's look at each of these core styles in detail.

Authoritarian

In this style, leaders tell their team what they want done and how they want it accomplished, without getting advice. Some appropriate conditions for this style in a project situation occur when you have all the information to solve the problem, you are short on time, and your team is well motivated. Many might think this style is based on yelling, bossing people around, and being demeaning, which is more a case of abusing power and should not be in a leader's repertoire. Authoritarianism should be used on rare occasions. To get things done, you need the other two styles.

Participative

Here, the leader includes one or more of the team in the decision-making process while maintaining the final decision-making authority. Using this style is a sign of strength, and the team will gain respect for you. It is typically used when you have part of the information, and the team has other parts.

Leaders can't and won't know everything, nor should they be expected to, which is why you have team members with the skills and knowledge to get the work done. Being a participative leader has mutual benefits—it strengthens and bonds the team and allows better decisions, particularly on projects where core subject matter experts (such as technology resources) might have key information, and understanding this style really can draw your team to you and drive a high level of results.

Delegating

Here, the leader allows the team to make the decisions. However, the leader is still responsible for the decisions made. This is used when teams can analyze the situation and determine what needs to be done and how to do it. You can't do everything! You must set priorities and delegate some tasks. This is not a leadership style where you can blame others when things go wrong. You must fully trust and have confidence in the team. Don't be afraid to use it; just use it wisely. In a project environment with stress from tight deadlines, delegating can be challenging, as there is risk of failure. Pick the right situations for this style; you must be willing to accept things not going quite as planned.

A Continuum of Leadership Style

Another approach is to consider style from a task as opposed to people perspective. Task-driven leaders focus much more on the work, whereas people-driven leaders consider the needs of people in making decisions and driving project results. I particularly like this way of considering leadership. It frames the type of project manager we might have originally been and our growth over time. It also considers the challenge of being a project leader where tasks and people often conflict. Looking at our leadership style from a situational perspective might guide how we dial up or down various parts of who we are as a leader.

The Leadership Grid, also known as the Blake Mouton Grid (Figure 4.2), plots the degree of task-centeredness as opposed to people-centeredness and identifies five combinations as distinct leadership styles. DiSC® personality profiling uses the same concept.

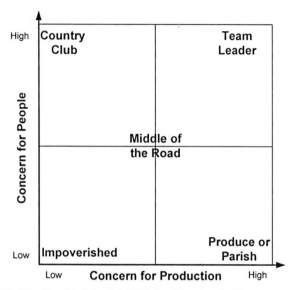

Figure 4.2: The Leadership Grid (Blake Mouton Grid).

Some leaders simply want to get things done. Others want people to be happy. And others are a combination. If you prefer to lead by setting and enforcing tight schedules, you tend to be more production- or task-oriented. If you make people your priority and try to accommodate team needs, then you're more people-oriented.

Neither preference is right or wrong. It is useful to understand your natural leadership tendencies so you can begin to work on developing skills that might be missing. Read the different styles and follow the following steps I have outlined.

Using the axis to plot leadership "concerns for production" as opposed to "concerns for people," the Leadership Grid defines five leadership styles:

Country Club Leadership—High People/Low Production

Here, the leader is most concerned about the needs and feelings of team members. This leader operates under the assumption that, as long as team members are happy and secure, they will work hard. What tends to result is a very relaxed and fun work environment where production suffers because of lack of direction and control.

Produce or Perish Leadership—High Production/Low People

Here, people believe teams are simply a means to an end. Team needs are always secondary to the need for efficient and productive workplaces. This leader is very autocratic, has strict work rules, policies, and procedures, and views punishment as the most effective means to motivate the team.

Impoverished Leadership—Low Production/Low People

This leader is mostly ineffective. He or she has neither a high regard for creating systems for getting the job done, nor for creating a satisfying and motivating work environment. The result is a place of disorganization, dissatisfaction, and disharmony.

Middle-of-the-Road Leadership—Medium Production/ Medium People

This style is a balance of the two competing concerns. At first, it might seem an ideal compromise. When you compromise, you necessarily give away a bit of each concern so neither production nor team needs are fully met. Leaders using this style settle for average performance and often believe this is the most anyone can expect.

Team Leadership—High Production/High People

This is the pinnacle of managerial style. These leaders stress production needs and people needs equally. The premise is team members are involved in understanding organizational purpose and determining production needs. When teams are committed to and have a stake in the organization's success, their needs and production needs coincide, creating a team environment based on trust and respect that leads to high satisfaction and motivation and, as a result, high production.

Use the grid to mature your style

1. Identify your preferred style.

    a. Think of situations where you were the leader.

    b. For each of these situations, place yourself in the grid according to where you believe you fit.

2. Find areas of improvement to develop your leadership skills

    a. Consider current leadership approach; critically analyze effectiveness.

    b. Look at ways to improve. Are you settling for less because it is easier?

    c. Identify ways to get skills and be more effective to reach the team leadership position.

    d. Check your performance. Observe when you might be slipping back to old habits.

3. Put the grid in context.

The Team Leadership style isn't the most effective approach in every situation. For example, if your project involves a transformation or other significant change, higher emphasis on people than on production might be needed. Similarly, when faced with an economic hardship or physical risk, people concerns might be placed on the back burner.

Unique Ability Approach

The Unique Ability approach (as developed by Dan Sullivan of *The Strategic Coach*—Figure 4.3) has immeasurably helped me know my style. It is founded on asking others against a 2x2 grid what our capabilities are.

- **Incompetent:** Don't ever ask me to do this. I really can't do it, and I would be a hazard to those around me (in sailing, don't even ask me to guide how we fix our engine!).

- **Competent:** I can do this kind of work, but many others are better than I am (spinnaker flying comes to mind).

- **Excellent:** I am excellent at these skills, and I could even help others develop these skills (racing tactics and navigational skills).

- **Unique:** I am passionate about these skills and, really, no one else has them on the team. Focusing on our unique abilities frees us to let others contribute with their unique abilities and enhance the leadership experience (anticipating the wind to come).

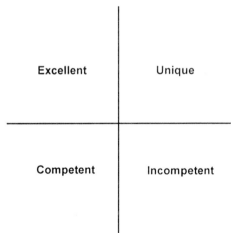

Figure 4.3: Find our Unique Ability.

This additional view frames our choices around what to lead, how to lead, when to lead, and when to let go. This type of reflection really allowed me to focus on my leadership style and when to apply it. One experience comes to mind. After knowing where to focus my abilities, I stopped worrying about every detail in the project. Although I was excellent at doing details, it didn't allow me to free myself to drive a higher-level result for the project. By giving responsibility for details to someone else on the team (who had that as his unique ability) I focused on my unique ability to "anticipate future needs before they happened," which in turn made a huge difference to the project's success.

The Five Degrees of Freedom

The final component I want to share is applying the Five Degrees of Freedom (Figure 4.4), which can help determine what you need to control and a level of trust with others. It, too, shapes our style.

Above the "freedom line" are the ways you can lead others to act freely. Degree 1 is the ultimate level of freedom—letting team members act on their own and report to you results of their action. Whereas, at the other end of the spectrum, Degree 5 below the line says wait until I tell you what to do.

The Five Degrees of Freedom help remind us to stay at the highest degree of freedom possible and provide freedom to others. In the end, if we all do our jobs and interact well together, any project has greater success.

1. ACT ON YOUR OWN—ROUTINE REPORTING ONLY

2. ACT—THEN ADVISE AT ONCE

3. RECOMMEND—THEN TAKE RESULTING ACTION

Freedom Line

4. ASK WHAT TO DO

5. WAIT UNTIL TOLD

Figure 4.4: The Five Degrees of Freedom.

The Value of Your Style—Creating Results

Being a leader is more than just managing and guiding. It is difficult work—because, in the end, it is really about people and getting work done through them. Knowing your style, learning how to adapt and apply the right amounts at the right time is the art we so often talk about in the project world. It needs daily exercise, practice, and application.

The real value of knowing our leadership style is in the ability to:

1. Create clarity, direction, purpose.

2. Bring calm to tense situations.

3. Have positive engagement of divergent teams, cultures, and environments.

4. Recognize it isn't just about problem solving. Rather, it is more about dilemmas and choices to be made, and as a leader, how we take advantage of or create opportunities to do things differently.

The Keys to Our Success

In sailing, as in the project world, you can't get things done without people. Guiding a sailboat through open water that might contain troubled waters requires the application of adaptive and creative leadership. Knowing our leadership style affects the result—arrival at our destination with the boat and crew intact.

Catherine Daw, MBA, PMP, CMC

Catherine is president and cofounder of SPM Group Ltd. She provides the vision and leadership needed to evolve the firm, including the current corporate direction in strategic initiative management. Her focus is what matters most to clients— solutions that exceed expectations, save time and money, and help achieve superior business results.

Catherine is a pioneer with more than twenty-five years of experience working in a variety of public- and private-sector companies, holding progressively higher positions both in information technology and business areas. Catherine's personal idiom has always been to deliver results on a foundation of integrity and open communication with a strong understanding of value creation and strategic advantage for businesses.

Catherine holds a Bachelor of Science degree in Mathematics and Computer Science from Queen's University and a Masters of Business Administration from York University. She is an active member of the Project Management Institute, a certified Project Management Professional (PMP), and a Certified Management Consultant (CMC).

As a renowned expert, Catherine is a frequently sought contributor for her views who delivers presentations in Canada and internationally. She is a regular columnist for many publications and blogs, frequently on core topics for which she is passionate.

Catherine Daw, MBA, PMP, CMC

President, SPM Group Ltd
1200 Sheppard Avenue East, Suite LL02
Toronto, Ontario
Canada, M2K 2S5

E-mail: Catherine.Daw@spmgroup.ca

Website: www.strategicinitiativemanagement.com

LinkedIn: ca.linkedin.com/in/catherinedaw/

Twitter: @Catherine_Daw

Professional Skills

The Keys to Our Success

CHAPTER 5

Good People Make Super Project Managers

By Derek Vigar

I have a lucky T-shirt. It's well worn, made from the most comfortable of cottons, and sports a faded iconic Superman crest on the chest. My trusty T-shirt has served me well over the years, and sometimes, it secretly gets tucked under a dress shirt when a bit of extra reassurance is needed for an important lecture or executive meeting. I like to wear it because it feels good and likely subconsciously reminds me of the superpowers I can summon to be a superhero to my students, clients, family, and friends. Although it occasionally attracts some good-spirited jeers for being a tad bit too form fitting, the way I see it, a well-loved vintage tee is still way more socially acceptable to wear these days than tights and a cape.

Superman to the Rescue

When I think about the key that differentiates those who stand out in the profession, the "Man of Steel" lends inspiration. Yes, the whole super-strength, invulnerability, superior vision, acute hearing, faster than a speeding bullet "thing" is admittedly impressive. I'm envious alone of how he can hop in a phone booth and be ready for work in a matter of seconds considering the superpowers needed at our house every morning to get everyone out the door. But for me, what makes him truly an extraordinary hero is that he balances his super abilities with a strong moral code. We love Superman not only because he has super powers, but also because he is a super person.

> **My faithful Superman shirt reminds me that to make a real difference, super abilities in project management tools, techniques, and technologies need to be balanced with a strong sense of being trustworthy and genuine.**

Traits such as honesty, idealism, and integrity can lead you to being *faster* at delivering projects, *more powerful* as a trusted advisor, and *able to leap* through projects with credibility and respect.

Trust Is the Key

As a college professor, I work every day with young professionals motivated to learn the difference between being good and being great. The Jenga® tower of textbooks stacked above my desk reveals an alphabet of project management topics ranging in everything from Agile to Zero Float. Although all this content helps shape the expertise I bring to clients or the classroom, you won't find the "It Factor" that distinguishes me as a professor and consulting professional on any of those pages.

Taking a gulp and asking clients, partners, and students for their insights (which in one case was greeted with "Why do you want to know... Are you dying?") yielded a surprising chorus of comments that rallied around a common theme: I had earned respect both as a professional *and* as a person.

I had developed credibility by combining project management subject matter with personal substance. They wanted me on their team, no matter the project specifics, because there was earned confidence and trust that I would do a great job. I had disproved that adage "It's not personal. It's just business" by showing what macho-manly-men like me aren't supposed to share... I cared.

Some of my students possess strong study acumen and can absorb a tremendous amount of project management content. Yet, those who seek me as a mentor do so because they recognize that taking a huge step forward in their career is so much about gaining experience and confidence with the people side of the profession.

- How do I get people to call me back?
- How do I facilitate a boardroom full of competing interests?
- How can I negotiate up a deliverable on a team member's priority list?
- How can I get the team to "buy in" earlier and generate momentum?
- How do I develop strong relationships leading to continued work opportunities?

I offer that *the key* to many of these questions lies in being credible and earning trust. Subject-matter expertise has a role, but your commitment to approaching the project every day with honesty, trustworthiness, and integrity is just as important a predictor of project and personal success.

> **Striving to be a good person can be the best path to being a great project manager.**

"C'mon, You're Kidding Me, Right?"

I know; I know. You paid good money for this book expecting some secret inside scoop, and all I'm giving you is a modified Boy Scouts motto. Well, let me show you the tangible importance of my claim, and to do so, I'll call on Jim Kouzes and Barry Posner, who have done extensive research focusing on what makes "exemplary leadership."[1]

By analyzing thousands of personal best leadership experiences, they derived their popular "Five Practices of Exemplary Leadership,"[2] and ultimately concluded "credibility is the foundation of leadership."[3] The authors administered a questionnaire to more than 75,000 people worldwide asking, "What do you look for and admire in a leader?"[4]

So, what do you think they found was the number one characteristic?

- The ability to recite project management inputs and outputs?

- The power to create a Work Breakdown Structure while blindfolded?

- Or more seriously, was it the ability to deliver against our profession's notorious mantra of "on time, on budget"?

---

[1] Kouzes, James M. and Posner, Barry Z. *The Leadership Challenge: How to Make Extraordinary Things Happen in Organizations.* 5th ed. San Francisco, CA: Jossey-Bass, 2012.

[2] http://www.leadershipchallenge.com/About-section-Our-Approach.aspx

[3] http://leadershipchallenge.typepad.com/leadership_challenge/2009/03/credibility-is-the-foundation.html

[4] www.hbr.org/2009/01/to-lead-create-a-shared-vision/ar/1

Well, the trait that consistently rose to the top across many cultures, receiving more than 60 percent of the votes, and surfaced as the most important characteristic people looked for in a leader was...

Honesty

No, really, it's the truth. Further, when Kouzes and Posner researched attributes people most admired in their colleagues, honesty again topped the list.[1] I guess all this research confirms that, although our profession is rife with processes and procedures, honesty really is "the best policy."

Being honest with clients, sponsors, teams and vendors is critical for building trust and credibility. No matter the industry, people insist that their project managers be truthful, principled, and ethical. People inherently respect those who have confidence acting with integrity and know where they stand on matters of importance.

"Be Good"

It sounds so simple. "Be good" is a message we've heard since we were kids squirming at the dinner table or during long family drives when we couldn't keep our hands to ourselves in the back seat. During our lives, it's been emphasized countless times. So why is it not always top of mind on projects?

The reality is the short-term nature of projects amplifies the pressures to perform. Status reports ask repeatedly "How are we doing?" against estimated goals set for budget and timeline. The answer to that question invariably influences post-project performance reviews and perceptions of success. With this, it takes considerable courage, especially for people early in their career, to push for what they perceive as a necessary change when it has the potential to cause a delay or require extra investment.

Each semester, I ask my project management students to visualize what, in their minds, would be the worst project. It's fun in a classroom setting because students invariably become competitive and try to outdo each other with scenarios that grow progressively more horrid.

One student starts us off with "I could never take a project in a slaughterhouse." (The class gasps.) Another chimes, "Could you imagine a project treating raw sewage?" (The class "ewwws.") Then, a lone voice from the back of the auditorium injects, "There's no way I would ever work for the Leafs."[5] (The class erupts.)

After all students have shared their worst-case scenario, I tell them that, coincidentally, someone in my network is hiring for that very project they proclaimed to distaste. The pay would be about $40,000, and if they wanted, I could put their name forward. About a quarter of the students put up their hands. I then raise the salary to $60,000 and get a few more takers. An $80,000 offer usually has three-quarters of the class engaged, with the rest seemingly holding off to see whether a higher offer will come. When one does at $100,000, typically 90% of the class is in. That means vegans have accepted projects at meat production facilities, and kids whose parents were affected by lung cancer wanted to work for agencies representing large cigarette manufacturers. I'm not sharing so you pass judgment, but to emphasize how tempting it is for people, especially early in their careers, to put aside their personal values and beliefs in pursuit of what they perceive necessary to be "successful" in the profession.

In a business environment where organizational loyalty is decreasing, personal values often are quite different from organizational ones. On top of this, stakeholders with competing interests often pull the project manager in different directions. There can be a perception that if you're not willing to do something, there's someone right behind you who is. Young project managers can rationalize non-ideal behavior because "it's only temporary," and when the project ends, they invariably will move on to something else.

Although it might create friction, the key is to recognize the importance of consistently doing both ***things right*** and the ***right thing***. Championing the best interests of all project stakeholders will likely instill the kind of trust and respect that influences success. Projects will always start and stop, but your credibility, reputation, and personal brand will last a lifetime. As social media blurs the line between the person and the professional, integrity and trust have never been more important anchors for your brand.

Soft Skills Lead to Hard Results

The risk in packing this chapter with words such as **honesty, integrity** and **trust**, is that one might assume this is only a warm and fuzzy *Chicken Soup for the Project Manager's Soul* tale. However, I wanted to share with you two examples of how these keys have led to tangible and differentiating results for my practice.

---

[5] One day, my hope is that someone will pick up this book and not get this reference. At the time of publishing, our beloved Toronto Maple Leafs have gone seven seasons without playing a single playoff hockey game (the most of any team). Their last championship was in 1967, making their current forty-five-year drought between titles the league's longest. If the team's fortunes ever change, and there's an unfathomable Stanley Cup parade in Toronto during my lifetime, I'll gladly accept that this reference might no longer be relevant.

The faster you build trust, the more quickly we progress.

We have all been in a project situation where we wished we had more time. Most of us are working off the side of our desks, which means however good our intentions, we often can't dramatically ramp up effort until a deadline approaches. Figure 5.1 not only depicts this uphill relationship between effort and time, but also highlights a phenomenon that most of us know all too well— procrastination. This curve is the reality of both today's project environment and human nature (and likely why after coordinating twenty-four other submissions for this book, I'll be the last to submit mine).

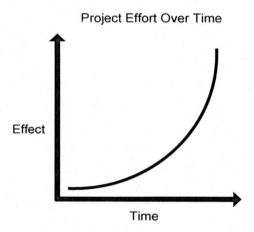

Project Effort Over Time

Figure 5.1: Effort ramps up as the project delivery date approaches.

The challenge with this type of workflow is that it leaves little room for error. Burnout can be strenuous, quality can be pinched, and roller-coaster resourcing can make for forecasting fits.

One fundamental way we, as project managers, provide value is by helping to **get the ball rolling**. Since at the start, we're pushing "the ball" up this curve, we have to build momentum. To do this, we need the strength of many, and must engage stakeholders to secure quick wins.

> **Building credibility and trust with stakeholders is the key to enable the buy-in and momentum necessary in the early lifecycle stages to ignite more project progress.**

We can do this by:

- Being passionate about the project and advocating its value

- Setting professional expectations of how the project will be managed

- Demonstrating that you *really heard* and support stakeholder requirements

Using credibility and trust to ignite progress early in the project flattens the overall effort curve (Figure 5.2). The result is that instead of wishing for more time at the end of a project, we've made better use of time at the beginning. This extra project runway increases the likelihood of success because we have more time to implement proper quality assurance, manage risks and changes, as well as

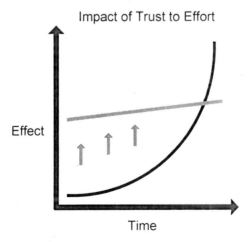

Figure 5.2: Trust helps foster buy-in and momentum for projects to progress earlier while flattening effort forecasts.

use our head start to hit delivery objectives. The investment we've made in people at the beginning will pay dividends during the project and beyond.

Your next project will likely come from a current advocate.

When I'm asked, "How do you find your projects?" the truth is they tend to find me. This statement by no means suggests that I'm passive or lucky, rather that I put emphasis on developing meaningful relationships with people who then advocate for me and my services.

A unique attribute of projects is that they are temporary and that people come together for a limited time to achieve shared objectives (Figure 5.3). When those objectives are accomplished, the project huddle "breaks," and stakeholders move on to other initiatives (Figure 5.4). This universal law of projects represents a tremendous business development opportunity... if you can earn the trust of stakeholders during the limited window in which you work together.

Stakeholders "Huddle" to Deliver the Current Project

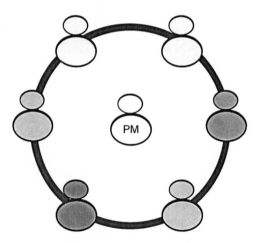

Figure 5.3: The PM has an opportunity to interact with a variety of stakeholders during the currect project.

Stakeholders "Break" and Move On to Their Next Initiative

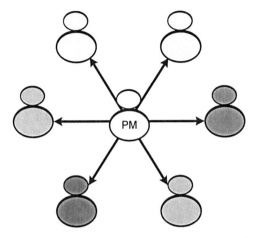

Figure 5.4: If the PM has made it a priority to build trust with stakeholders then they will advocate for the PM's services on their next engagement.

Consider the number of exciting engagements going on right now that involve someone who has previously seen you in action. What will these people need to remember about you in order to be your advocate?

> **If you want people to be eager to offer a *character* reference, you had better make sure that you seize every opportunity during the time you work together to show them your character.**

People will remember your character and the confidence they had in you more than the nuances of your methodologies, or even—dare I say it—whether the project came in slightly over budget. Treating interactions with project stakeholders essentially as **tryout** opportunities means that if you make it a priority to knock people's socks off, you will be memorable when these professionals move on to their next engagements.

I find it neat that when I look back over the past fifteen years of consulting projects, I can literally map connections between engagements and across industries, almost creating a family tree of how a team member, client, or sponsor from one project became an advocate leading to the next. This tactic wasn't conscious, but a by-product that resulted from the investment in people I felt was necessary, not only to deliver the current assignment, but also to really enjoy and draw meaning from the work.

Carpe PM—Seize the Trust

The benefits of building trust are limitless, just like the occasions that exist to earn it. The groundwork is laid through our daily interactions, decisions, and ethic, often at times when we do not ask or need anything in return. We often assume that our relationships are strong, but the reality is that the busyness of juggling projects sometimes makes it challenging to nurture them properly. (Think whether you've ever said, "How are you?" to someone you're passing and you ended up three steps past them before they answered.)

The project lifecycle presents many built-in events to develop trust, so be ready to fly into action and find opportunities to:

- Take ownership of an early deliverable—and then over-deliver on it
- Talk to stakeholders about their business goals and show how project requirements align
- Present "sugar-free" risks (aka no sugar coating) and proactively offer solutions to manage
- Share true status and highlight the project team's ability to handle adversity
- Dialogue with stakeholders about their honest perceptions of how things are going
- Grab onto a mentor you admire both as a person and a professional

Being genuine is both an art and a science, and imperative to developing trust. It's tricky to share a recipe for how to achieve it because the flavors and ingredients are so personal and different for everyone. Folks I consider "real" seem to blend a hearty base of honesty and confidence with a mixture of sincerity and consideration; however, others add seasoning to the recipe by adding cups of reliability and dollops of personality.

Every holiday season, without fail, the first card to come is from a real estate agent who I assume got our name from an Open House sign-up sheet years ago. Now, I'm sure there is a business textbook out there somewhere recommending this as a practice for "being personal" and building rapport with prospective clients, but at the risk of sounding like a Scrooge, the gesture lacks substance because the relationship was never developed.

Now, compare this with my new friend and book partner, David Barrett, who called me during the holidays to tell me that he was on his way over. He's a super busy guy, but he made it a priority to come by in person and drop off a holiday basket spilling with treats and goodies. His visit was brief, but in that time, I felt he was genuinely interested in learning more about my family, my home, and how I look when I answer the door in my pajamas. I mention this not only to encourage David to continue to bring treats to our home, but also to highlight the fine line between doing what might be expected and successfully seizing an opportunity to be genuine and build trust.

You're Good; So Now It's Time to Be Super.

Look, I have a sneaking suspicion that you're already pretty good. My first hint was your obvious great taste in reading material, but on top of that, it's impressive that you're taking the initiative to connect and learn from others in the profession. If my hunch is true, then you already know a thing or two about "saving the day." You likely enjoy making things happen and have an "anything's possible" attitude. You're exactly the type of person who can use this profession to make a difference.

In 1940, Bud Collyer[6] became the radio personality who captivated audiences' imaginations with the iconic phrase "This looks like a job for Superman!" When you think about the values necessary to be successful in today's project management profession, I hope you couldn't agree more.

Honesty, trust, integrity, putting people first, and fighting for what's right are all differentiating traits that lead to success—not just in projects, but also throughout your professional career and personal life. Success will be a product of the credibility and personal brand you build by consistently doing both "things right" and "the right thing."

---

[6] http://en.wikipedia.org/wiki/Bud_Collyer

There will be challenges. There will be villains. There will be times you need to summon superhero strength to push back. However, that's what makes the story an adventure. And if your intentions are genuine and honest, you'll often find people will admire and respect your character.

It's difficult to find an actual phone booth these days, so it might not always be convenient for us as mild-mannered project managers to change into a superhero uniform (or vintage T-shirt).

But if we keep the values from this chapter **top of mind...**

It won't be as important that we wear them **across our chest...**

Because we'll know they are true **in our hearts.**

These are the keys to success. (And if they don't help, you're more than welcome to borrow my lucky Superman shirt.)

Derek Vigar

Derek is an award-winning professor and active project management consultant. As a faculty member with the School of Management, he leads the development and delivery of the Project Management Specialization at George Brown College in Toronto. Over the past twelve years, he has taught more than 5,000 students who have consistently shared through their course feedback that he is a dynamic speaker, passionate mentor, and deserving of one of the college's top teacher effectiveness ratings. In 2012, Derek was recognized from 3,500 employees to receive the College Achievement Award in Innovation, largely for his efforts in facilitating partnerships, field-education, and innovative learning for his students so they graduate with real applied project experiences.

As a freelance project manager, Derek has built a successful practice through commitment to the trust and integrity values highlighted in this chapter. He's provided project management leadership to more than forty different organizations, many of whom eagerly endorse him as both a person and professional. Derek is a compelling, hands-on consultant who invigorates projects with subject-matter expertise, people skills, and experiences drawn from managing projects of many flavors.

Derek is always looking for opportunities to be part of projects that make a difference, involve great people, and lend themselves to having a little fun along the way. He looks forward to connecting, so feel welcome to get in touch.

Derek Vigar

Professor, School of Management
Centre for Business, Arts and Design
George Brown College, Toronto, ON, Canada

Personal E-mail: derek.vigar@gmail.com

LinkedIn: ca.linkedin.com/in/derekvigar/

Twitter: @DerekVigar

Facebook: www.facebook.com/ProfDerekVigar

Phone: 416-415-5000 x3391

The Keys to Our Success

CHAPTER 6

Never Go in Alone

By David Barrett

Throughout my career running events, launching education programs, running small businesses, and more, the biggest lesson I have learned is that you cannot do it on your own. A partnership, a cooperative, an advisory board, or even just one advisor, a mentor, a coach... anything is better than just you are—and certainly more enjoyable!

Call it the power of two. Call it the power of many. Recite the famous poem "No Man Is an Island" (change the word "Man" to "Person"). They all mean the same thing.

As a project manager, we need people on our side—at the front end, during the project, and at the end. I need support, and I need advice. I need someone or many people to keep me in line and to help me along the way.

This book is a great example. I knew I could not do it alone so I pulled in my new friend Derek Vigar. The teamwork has undoubtedly made it easier, but as well, it has produced a very different product than anything I would have produced on my own. Moreover, it was much more fun working with Derek.

I have just completed another book called The Power of the Plan. Again, I knew I could not do it alone, so I pulled in another old friend, Doug Land. We worked on the project for more than two years (do not let anyone tell you that writing a book is easy!), and we could not have done it without each other. I don't go in alone.

In 1985, I left a very comfortable job at a bank and went out to develop custom software and sell computers to small businesses. A great idea back then and I was sure I could do it alone; instead, and very fortunately, I found Bob Icely, my future partner. He came with experience, contacts, and very important back then—dealership connections. Most important, he was someone with whom I could share the experience, which was important because it was not a great experience—very frightening, very stressful, and not very successful. I would not have lasted as long as I did without my partner.

One day I had an idea to create a training program for project managers that would be very different from anything of its kind in Canada. I had the vision, the time, and ability to do it, and I could make it work. However, very early, I realized that none of that was true. Therefore, I got help.

First, I took a course on creating a new business. It took a year, and I could apply everything I learned to this new dream as I went through the program. One instructor played a big role in my dream, as he reinforced the excitement and adventure of building something new such as a business.

When I was about to launch, I made the best decision of my whole life (OK, the second best after marrying my wife)—I asked a major Canadian university to partner with me, adding potential exposure, expertise, and resources. The Masters Certificate in Project Management at the Schulich Executive Centre was born. I didn't try to do it alone.

A business partner, a transaction partner, or a partner of any sort will provide many parts of the puzzle that will be missing if you try to go in alone. Most of us cannot do it all. One hesitation is that we typically give up half the profits! However, the truth is half X is often much better than all Y or better than half nothing. Most often, if you look back, you will realize that the partnership could take the company, idea, or business further than you could have on your own.

My wife Karen and I run many projects on the side—a travel website, a site for project managers and not-for-profits to connect, events at our church (annual auction fundraiser, progressive dinner event), our ski club (dances, fundraiser, our cottage area art show and more). Everything we do involves a team of people, an advisory group, and others.

Running the auction at church is about the team, but this is such a difficult project to run for so many reasons: politics, volunteers as opposed to full-time staff, unqualified helpers, and more. This is a perfect example of a great time to have a codirector or coordinator. You can get yourself into lots of hot water on a project such as this. Find someone to share the heat, the stress, and the work. Ideally, find someone different from you—with skills that complement yours. You do not need to be best friends. We can never do it alone.

As project managers, we all know the power of the team—good leadership, strong communications. I am not talking about this. I am talking about a different relationship, a different team.

What I have learned from the start of all this is that although you need a good team to execute the project, it is just as important to have help and/or guidance in the role as a project manager. You need an advisory board, a coach, a mentor, or a confidant.

If I worked in a large corporation and managed a large, mission-critical project, I would seek someone with whom I could sit once a month. Before I selected the person, I would identify the potential issues I might face during the project—politics, leadership or team issues, budget problems. Then, I would short-list the people approachable, knowledgeable, and possibly strong in the identified soft spot. He or she would be senior to me and more experienced. The truth is that most people whom you would ask to do this would be honored and thrilled. If they are senior enough, they probably will have a stake in your success and that of your project.

Your "ask" must be framed properly. It must be a structured relationship with a full commitment from both parties. It must guarantee accountability—especially in your role.

The accountability issue is big to me. I am often asked for advice, especially from young people, on careers, work issues, and more. I typically start these meetings with a promise of accountability. If "we" decide that a course of action is good—he or she is accountable to me to get it done or follow up about why it was not the chosen course of action. Sounds funny, but I just don't like wasting my time giving advice and not hearing back.

So, find a senior person and frame the request well so it is understandable and so the expectations are clear. Commit, and go for it. The meetings do not need to be half-day affairs. They should be an hour at least. Like all good meetings, they must start on time and end on time. If you needed more time—make the next meeting longer. You are both too busy to wing it.

If I were working on a project for a smaller business without the required mentor, coach, or advisor within, I would find that person outside the company—someone who, again, has been through the same type of project in the same type of environment. Same structure, same approach.

In my life, I manage many projects—— conferences, training programs and more—that are delivered to a large community of people. My advice, my coaches/mentors, comes from an advisory board. Here is how I work this.

I am about to launch a new event in a major US city. Although I haven't completely committed to the event yet, my research suggests that this is a good idea. My first step is to gain support of the local community. I start to pull together an advisory board at this time. I find the key players in the space I am working (typically project management or business analysis). I invite them to a dinner gathering to discuss the idea. This is a relatively inexpensive way to gather good information in a short period. I pay for everything.

Out of this dinner, first comes a formal commitment on our part to move forward. Then, I formally create my advisory board. It might be the same group I met or more likely something a little different. Sometimes, I realize that some people I have selected are not a good fit. This is the opportunity to deselect them. OK, deselecting someone is the toughest part of all of this. Ouch! Very uncomfortable and very difficult for all. However, I am afraid you have to do it. For me, I just communicate honestly. "Thank you for your offer to help, but for his round, we are well looked after." Still ouch.

I call another more formal meeting of the new advisory board, and I go to work. I use these folks to help me understand market and business trends, to identify topics for the conference, and to help with content selection. This group also provides a good "visual" for the event in the eyes of the local community. They want to see their peers in a position of influence in a local event.

We meet twice a year for three hours over a working dinner in a great restaurant. This approach is inclusive for the community, comforting for me, good for the business, and great for our customers. I never try to do it alone.

More and more these days, I hear of people hiring life coaches. It is not a new idea, but one that seems to be catching on fast. Expensive, but very valuable. A life coach will help you through any aspect of, typically, your professional life, but he or she can certainly touch on your personal life and other aspects. Life isn't easy, and in truth, if you have particular goals in mind, you could use a coach. Many of us could use this sort of coach. It is easier than going it alone.

So, my lesson learned and advice to anyone running any type of project, including your life, is to get some help. Find a partner, coach, mentor, or advisory board. Get them engaged in your plans and then pick their brains. Pick people or one person who complements you, but isn't just a "yes person" for you. Find someone whose company you will enjoy and who will challenge you. Find a group or someone smarter and wiser than you are. Find someone who you will want to stand beside in the end and say, "We did it... together." Well worth it. I promise.

David Barrett

David has spent the past eighteen years building a series of companies dedicated to the business of project management:

Running conferences in Canada, the US, Australia, and India

Running education programs at ten universities across Canada

Operating one of the world's largest online portals for project managers

He is also the director of PMPeople.com, a website connecting project managers donating their time to charities and not-for-profit organizations across Canada. David coauthored *The Power of the Plan: Empowering the Leader in You.*

David and his wife Karen run an online travel business called SharingTravelIdeas.com. He has four children ranging from seventeen to thirty, and he will soon be a grandfather. He is a skier, a woodworker, a beginning golfer, and a road biker who hates hills.

David Barrett

Diversified Business Communications Canada
ProjectWorld * BusinessAnalystWorld
*   ProjectSummit * APC Canada * Canadian
Leadership Summit

Schulich Executive Education Centre
Schulich School of Business
York University, Toronto

ProjectTimes.com; PMPeople.com;
SharingTravelIdeas.com

E-mail: dbarrett@solutionsnetwork.com

Website: www.DavidBarrett.ca

LinkedIn: David Barrett

Twitter: dbarrett1

CHAPTER 7

You Can't Phone It In!

By Sandra Hoskins

The most consistent lesson I have learned in project management is that "You can't phone it in. If you want to carry the title, you have to do both the project management and project work. You'll note that I didn't say that you had to like the work, but you have to do the work."

When you are assigned as a project manager, you become the lynch pin for stakeholders who have a specific end-game in mind. They might not articulate their goal, vision, or product exactly the way you need, but it is up to you and your team to figure it out.

For many project managers, a project starts as a great challenge with great potential. Then, a second project is added to the portfolio, then a third, then another and another. Project management moves from a passion to a job. When that happens, it becomes difficult to remember that a project manager's role is to protect the team instead of filling the days with endless meetings and conference calls.

It doesn't help when the project manager is also an active team member. It leads to conflicting roles. The project management role becomes less important over time, and project management deliverables are relegated to the back burner. Which is more important—the project work or the project management work? The short answer is "they are both important."

When you accept your role as a project manager, it doesn't matter what the type of project is, you become accountable for ensuring that the basic principles of project management are applied proactively and consistently. Failing to apply these principles can negatively affect the relationship between the project team and the stakeholders. I have often heard the mantra "the project is too small for a schedule." The message sent to your team members and other project stakeholders is that a schedule is only required when the project is big enough, important enough, costs enough, or is visible enough. There are other excuses, but these are the heavy hitters.

I have been fortunate enough to work on projects in many industries. At a minimum, the project manager must ensure that the stakeholders have access to the right information at the right time to make the right decisions. I have been fortunate enough to learn that you can be a successful project manager using a "really lean approach" to project management deliverables. In our organization, we refer to the lean process as the Scope—Schedule—Results—Leadership (SSRL)™ framework (see Figure 7.1).

The SSRL Framework

As a project manager, business, and entrepreneur, I have found that the simpler the process the easier it is to implement. After working on a variety of projects, I have learned that regardless of scope, size, cost, resources, or location, I need the fundamental four: a scope statement, a schedule, accurate results, and leadership skills. Every project needs a starting point. The scope statement is that starting point.

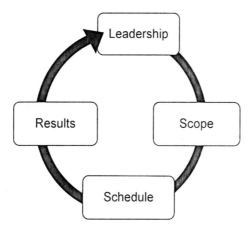

Figure 7.1: The Scope—Schedule—Results—Leadership (SSRL)™ framework.

Creating a Scope Statement

Every project has to start somewhere. You can create any number of documents that can trigger a project, but the bottom line is that without a scope of work, the stakeholders will never understand "What Does Done Look Like." When you define the scope of work required to complete the project, you create the basics of a relationship that allows the team to deliver against the stakeholders' needs and expectations.

A scope of work should be formatted to give the team three specific reference points.

1. The first reference point is to define *"what is in scope."* Many project teams do this but only for the deliverables that are "the responsibility of the team," bypassing the deliverables that are the responsibility of others. If you need to have inspectors or regulators approve your work, their deliverables and work must be included in the scope of work. When the stakeholder agrees to what is in scope, the team can begin the estimating, costing, resourcing, and scheduling of the project work.

95

2. The second reference point is to define *"what is out of scope."* It is critical to the team's success that the stakeholders understand and agree to the work excluded from the project. In many cases, we forget to reinforce that the work excluded from the scope won't be completed by the team and will not be estimated for time, cost, resource allocations, and risk. When excluded work is poorly communicated to project stakeholders, it becomes our biggest source of change. I have learned over time that it is as critical for the stakeholder to approve the excluded work as it is for them to approve included work. It gives you leverage when changes are requested and to manage their expectations better.

3. The third reference point is *"what is unknown but needs to be confirmed."* Every project team identifies work that could be included in the scope statement but needs to be confirmed by outside stakeholders. It is usually placed in a parking lot. When work is added to the parking lot, a team member should be assigned to follow and determine whether the work is required. If the work remains in the parking lot without a qualification or quantification, it becomes a source of risk to the project. It is imperative to address this work so the project manager can prevent "out-of-left-field" events that can delay the project and potentially cause cancellation.

When all is said and done, the **scope statement (a.k.a. the scope of work)** documents what product, service, or result the team is contracted to deliver to the customer. A well-written scope statement provides the team with a document confirming stakeholder expectations. Work considered in scope provides the framework for creating the work breakdown structure (WBS); estimates for time, costs, and resources; communication; risk; and the overall schedule.

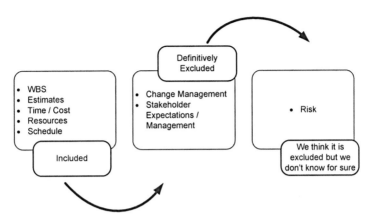

Figure 7.2: Building a scope statement.

Developing and Maintaining the Schedule

One of the first questions I ask project managers when I complete a forensic project assessment is how current is their schedule. I am often rewarded with a blank stare as they try to remember the last time the schedule was updated and whether it communicates accurately what work must be completed, when it must be completed, the scheduling constraints, the resource constraints, and whether the project is on time.

I have been fortunate to learn the "Fine Art of Scheduling" from a master. When I first started to work as a project manager, I had a habit of keeping the schedule in my head, only updating it when someone asked me for a report, and the percent complete button in Microsoft Project® was my friend. I also didn't use the schedule as an execution tool. I would complain when the schedule end date moved. (It was all Microsoft Project's® fault.) When I first started developing to develop schedules, the schedule was just boxes used to communicate where I thought we were, but not where we really were and what the future requirements were. It was a long six months turning me into a project manager that could run a project from a schedule.

Schedules are the roadmap for the work to be completed during the project. Without an accurate schedule that reflects real time estimates and real resource requirements, project managers are without the most fundamental tool for ensuring the stakeholders understand what is required, when, and who will do the work. Why would any project manager, myself included, put their team at risk? It is usually an oversight. The schedule takes time to maintain, and there is often little perceived value in a current schedule. The schedule is the only document that provides a common view of the critical path activities that must be completed to achieve the successful on-time delivery of the project. Use it accordingly.

Tracking and Reconciling Results

I have learned to think of tracking the actual performance or results as closing the loop. Too many project stakeholders are held hostage by bad data, which results in bad decisions. Team members should report results to the schedule a minimum of once a week. I have learned to be a great fan of reporting actuals to the schedule daily, but I am a realist.

If you ask people to report every day, it becomes one of those low-value activities relegated to the end of the week, anyway. If you ask for help to keep the schedule current, most team members find the time in the week to provide their actual work results. They need to trust that you will use the information to help facilitate their success.

• A schedule is only as accurate as the work results reported/submitted/contributed by the team members. A project manager must be able to reconcile the results at least once a week. I can hear the pushback now. But, I don't have the time. You are the project manager; you must make the time. If you cannot reconcile where you have been, you cannot forecast what must be accomplished.

Team members are always willing to contribute how much they have accomplished, how much time they need to complete their tasks, and who they might need to help. None of this happens without trust. Team members only tell you the way it is if they trust that you will accept their information without repercussions.

When you track results, it is what it is. You can only adjust the scope or the schedule based on the real-world results. Fabrication leads to communicating false expectations. You can only communicate what you know for sure.

Leading by Example

Project management is about accountability and leadership. When you are a project manager, one of the most important things you can communicate is your vision of what project management principles can contribute to the team's success. In other words, you must be able to give all stakeholders a vision of "what is in it for them." In order for people to follow you, they must believe that you are in their corner and will do everything needed to ensure their success.

Any person who has ever worked on a project, me included, has worked with a project manager who never had your back. If anything went wrong, it was your fault. The fact that there wasn't a well-defined scope of work, a schedule that reflected the project's reality, and a process for ensuring that the reports were accurate wasn't their fault. It was your fault, not the fault of the other team members, the stakeholders, and management. Somehow, it was never the project manager failing to manage the project. Project managers never lack responsibility, but they frequently lack accountability.

Sometimes, it is difficult to remember when the project isn't progressing as expected that it all starts with communicating the scope of work.

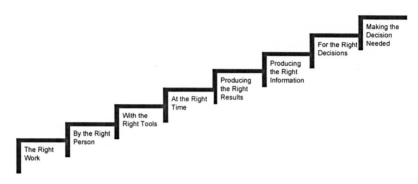

Figure 7.3: Steps to Effective Communication—Communication Start

In conclusion, I have learned, sometimes painfully, that project management is doing the work that needs to be done to ensure that your team can be successful under the most unpredictable of circumstances. To quote Aldous Huxley, "The only truly consistent human being is a corpse." The bottom line is that we work with people who are warm, vertical, and breathing. They can and do change their minds. If you aren't willing to do what it takes, then project management might not be the career for you. Our role as project managers is to do what needs to be done to protect our team and, by default, our stakeholders from the situations that set them up to fail.

We are obligated to ensure that the team has all the necessary information to deliver against our stakeholders' expectations. There are no shortcuts. There are processes that work better than others do, but we must sustain the vital four: scope, schedule, results, and leadership.

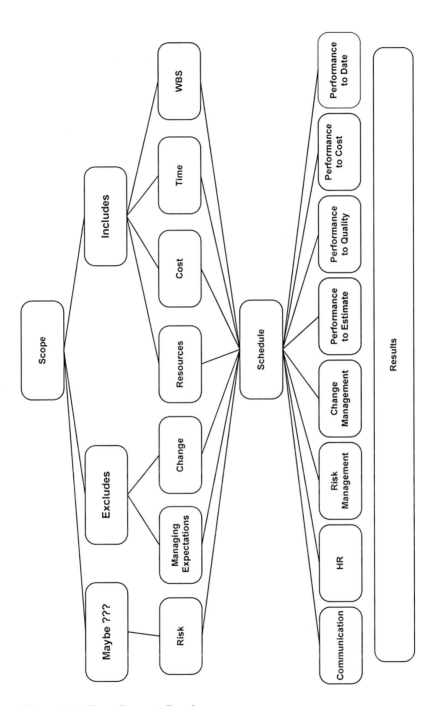

Figure 7.2: From Scope to Results

The bottom line: No one will give you permission to be a project manager; you are or you aren't. In the figure above, you can see the critical importance of a scope statement to launch your project, a schedule to forecast your project, and results to ensure that the team always knows what has been accomplished. You can't phone it in; you have to do the work. It is how we add value to the team, the stakeholders, and the organization.

Sandra Hoskins

As a speaker, mentor, educator, consultant, author, and project manager, Sandra is committed to working with people who want to create excellence in their life, whether at work, home, play, or in their community. Sandra's passions are project management and education. Her participation in these professions has taught her the importance of work-life balance.

A career spanning the military, business, public sector, and education has provided Sandra with a unique perspective on the value of project management and the practical application of knowledge, skills, and experience.

Sandra believes the principles of leadership, education, and project management are the difference makers and the deal breakers. It is how we grow people and organizations, and more important, it is how we influence lives. Project management cannot be an idea we simply talk about; project management is the action we must live.

Her personal motto is "project management is a life skill." When you think about it, project management is an approach to problem solving, no more and no less.

Sandra is the founder of The Kellan Group Inc. and a founding partner of The John Maxwell Team.

Sandra Hoskins, ISP, ITCP, PMP

The Kellan Group Inc.

The John Maxwell Team

Red River College of Applied Arts, Science and Technology

University of Wisconsin-Milwaukee

606-283 Portage Avenue
Winnipeg, MB
Canada, R3B 2B5

E-mail: sandra.hoskins@thekellangroup.com

Website: Sandra.hoskins@sandrahoskins.com

LinkedIn: Sandra Hoskins

Twitter: sjhoskins

Blog: www.pmalchemist.com

Skype: Sandra Hoskins (Manitoba Canada)

Phone: 204-253-1896
Fax: 204-254-6913
VOIP: 608-554-4616
Voice Mail: 608-301-5202

The Keys to Our Success

CHAPTER 8

The Upwardly Mobile Project Manager—You Can Get There from Here

By Kathryn Pottruff

I've noticed that being proficient at using standard project management processes doesn't always translate into stellar performance as a project manager. The adage that "soft skills matter" reverberates throughout organizations, but what does it mean, and how do people and organizations go about developing those skills?

Have you seen people promoted to a project management role because they are technically adept and then watched them flounder because they are inept at managing people? Project management is not for everyone; but if it is for you, then you'll want to develop the competencies that extend beyond just managing into true leadership. This chapter looks beyond the basics. It identifies the competencies and skills you need to develop if your intention is to lead project and program teams effectively.

I can't prove it scientifically, but I've witnessed a few trends at my client organizations. Rising-star project managers are promoted to senior leadership roles such as director, associate vice president, vice president, or even chief portfolio officer. I've noticed that effective PMOs are net exporters of talent.

How does this relate to the up-and-coming project and program managers? You can get there from here, but you'll need to develop specific competencies before the next rung on the ladder magically appears. You can enroll in courses to learn about project management processes, which will certainly help you on your way up the first few rungs of the ladder. From that point on, preparing for the next step becomes somewhat less obvious.

The skills you need to develop aren't typically acquired in classrooms; they tend to develop through on-the-job experience under the watchful eye of a successful mentor. The challenge lies in finding a mentor with the necessary experience who is willing to spend time helping you discover and enhance your strengths. In the rest of this article, I've highlighted competencies you should develop. I've also included suggestions on how to build these competencies.

Know the Company

If you want to ascend the corporate ladder, understanding the business and its operations is essential. At one point in my career, my boss told me that until I had five years in, there was little chance of my becoming a project manager. He viewed "feet on the ground time," as a prerequisite for understanding the culture and establishing a reputation that would help me build a useful network. People hesitate to share pertinent information with you when you haven't been there long enough to build and nurture trusting relationships. Later, that manager became a mentor of sorts and introduced me to people who positively influenced my career.

A program or portfolio manager has to know the business to be effective. To keep a program visible at the highest levels, a manager needs to know how the work supports broad organizational goals and strategic objectives. This insight is helpful when you need to communicate a program's value to other program managers or project managers and their teams. Without purpose, there is no buy-in.

All projects and programs involve a modicum of change. Business knowledge develops over time, and that knowledge helps a program manager prepare the business for change. Change requires advocates willing to communicate the benefits of the change well before implementation. Your network of contacts will be invaluable in communicating and supporting the change your program will ultimately bring.

Suggestions

- Keep your eyes peeled for potential mentors in the organization and keep in mind that the best mentors don't have to come from your workstream. A supply-chain project manager can benefit by having a mentor in accounting, finance, or manufacturing. Each provides a different perspective on how a supply chain project affects their area.

- Establish a relationship with your sponsor. Ask questions about how your project supports organizational goals. Schedule a fifteen-minute meeting and tell him or her in advance that you would like to understand the strategic direction of the organization. You might be surprised by the response. No one might have asked this before, and by initiating the conversation, you have just increased your visibility. As a consultant, I've been taken aback several times when I've posed these questions of my clients. Frequently, strategy is "under development," or as one vice president told me, "Our strategy isn't written formally; it's rather flexible."

- When it comes to managing change, be prepared to communicate openly and frequently with your key stakeholders. Listen to their concerns, and let them know as much as you can in advance. Reassure them that any transition will be smooth and supported. Invite them to ask questions and then answer them fully.

Be a Quick Study

How quickly can you adapt to new technology or ambiguous situations? If you are in the middle of a project, and there are many critical unknowns, are you paralyzed by anxiety, or can you take the situation in stride and devise a solution? Take a logical approach to solving the challenges that arise and consider what you've learned in the past. Those who can progress despite unknowns are among the most valued. Whining over unknowns accomplishes nothing and tends to make you overlook opportunities staring you right in the face.

Suggestions

- Train your brain! Learning agility is improved by doing puzzles; crosswords or Sudoku work well. If you're serious about this, try the brain training websites such as Lumosity.com or Neuroactive.ca. These sites help you deal with ambiguity and keep your brain functioning at its best.

Make Great Calls

Can you make good decisions in difficult situations? People make situations more difficult than they need to be. Good decisions are timely, appropriate to your level of authority, and they align with the organization's risk profile.

We each make decisions every day. Portfolio managers decide which projects are of the highest priority and how organizational resources, both capital and human, are allocated. Program and project managers decide how to prioritize projects and deliverables. Personally, we each decide how to spend the most precious resource of all—our discretionary time.

Suggestions

- Don't be overwhelmed by complexity. Compartmentalize issues, and you'll discover that when several small problems are dealt with effectively, the big problems become more manageable.

- If you make complex portfolio decisions, use a transparent and rational decision-making process that makes your decisions easy to defend and communicate. Learn about and use software tools that guide you through the prioritization process using clearly defined rating scales. Portfolio software can synthesize huge volumes of information and provide clear answers that would otherwise be beyond our reach.

- With discretionary time decisions, make your family and the time you spend with them your top priority.

Don't Give Up; Don't Give In

As a project manager, you must demonstrate ownership and commitment to success daily. Be passionate about managing the team and the work while ensuring that you meet internal deadlines. Team members appreciate working for project managers who hold themselves accountable for the project's success and who strive to create an open and considerate working environment. Take time to establish and maintain relationships with your team; encourage each person to do his or her best work.

Effective project managers are realists who exhibit strong self-control and self-reliance. Calm and able to respond appropriately in the most demanding situations, they never lose sight of the desired project outcomes.

Suggestions

- Busy, effective people are always top of mind when it comes to finding someone to manage any piece of work. If you've proved yourself as a top-performing project manager, you'll be asked to take on more work. You are likely to hear phrases such as, "It's not much work, but I need someone reliable to make sure it gets done right." If you intend to stay committed to keeping your project on track, you'll need to learn to say no loudly and clearly. Of course, it's not always possible or wise to refuse to take on more work, but you need to know your limits and stick to them. Think boundaries.

Take a Chance, but Think about It First

Every project presents with elements of risk and opportunity. Being able to identify, assess, analyze, respond to, and document risks throughout the project life cycle are important skills. I have worked with project managers who seemingly have a sixth sense when it comes to risk. They seem to understand intuitively project interdependencies and can minimize the impact of unforeseen changes to project outcomes. Not only are they aware of project specific risks, they can also make them visible to appropriate stakeholders and at the appropriate level in the organization.

Knowing how likely it is that a risk will occur and what its likely impact is are skills that develop with experience. Those who don't have experience yet would be wise to find a mentor who can guide them through the unknowns.

Suggestions

- Get your team involved in identifying, assessing, and analyzing risks early in the project, and then repeat the risk processes regularly.

- Keep your eye on issues, constraints, and assumptions, as these can become risks in a moment's notice.

- Create a risk register, and keep it up to date. Assign a team member as a risk manager for each risk.

- Risks and risk updates should be discussed at each status meeting.

- Don't become paralyzed in gathering and analyzing risk information. Adequate information is a good thing, but there is a right time to make a decision, and there is a right time to act. Sitting on the fence can be as dangerous as making a wrong decision.

Garnering Stakeholder Support

Project managers must know who their stakeholders are and what each wants from the project. Establishing partnerships with each stakeholder is imperative to success. Internally, the project manager liaises with the project sponsor, functional managers, team members, other project managers, and subject matter experts. Each stakeholder has his or her agenda, and unfortunately, it is the norm for these agendas to be in conflict.

It is up to the project manager to surface the conflict and discuss it openly with all involved until some form or resolution is achieved. Externally, the project manager is in touch with customers and suppliers, and he or she might have to communicate with unions, the media, regulatory bodies, government agencies, lobbyists, and so on. Knowing "what" is important to "whom" is paramount in managing stakeholder expectations and in garnering their support.

Suggestions

- Early in the project, conduct a stakeholder scan to gain understanding each stakeholder's objectives. Use a four-quadrant chart that depicts the stakeholders' power and influence with project outcomes, and then document the information in a stakeholder register. Managing stakeholder expectations is an ongoing activity throughout the project. Stakeholders have been known to change their perspectives and expectations as the project progresses and more information becomes available. The best way to manage stakeholders is to establish a solid relationship with them early, and then communicate with them frequently.

Leading the Team

Leading a team is perhaps the most challenging item for a project manager. First, you need to understand the competencies and skills each team member brings. It's helpful if you can get the team to share what they like and dislike doing. This isn't a promise to accommodate everyone's wish list, but you can't assign the right people to the right tasks if you have no inkling of skills, likes, and dislikes. Beyond existing preferences, it is useful to know whether team members are interested in developing new skills.

Effective project managers are willing to share their knowledge and expertise with others involved in the project. When appropriate, the project manager should share information about the project's alignment with strategy. In this way, the project manager can motivate the team in supporting organizational goals.

Most important, the project manager should strive to create a positive working environment where trust and good morale abound. To this end, the project manager must build relationships with the team. You should celebrate individual and team successes and learn to work collaboratively to solve the inevitable issues that arise.

Suggestions

- Communicate openly with your team. Encourage participation, and learn to listen. You have two ears and one mouth. Use them proportionally.

- At the Project Kick-Off meeting, draft a team charter that outlines how the team will work together. This exercise establishes the rules of engagement. It's easier to have this conversation before the work starts and issues arise. When everyone understands the ground rules, there tend to be fewer issues with which to deal.

113

- Team-building activities can establish relationships outside the work environment. When team members know what they share, work runs more smoothly.

Know Yourself

Be comfortable in your skin. Build a strong foundation, so you aren't buffeted by the vagaries of organizational life. Figure out what your strengths are and what makes you happy. What do you feel passionate about; what you would do every day, even if you didn't have to? Sometimes, the answer to the question is elusive, and other times, you just know. The closer you can align your work with your strengths and passions, the happier you will be.

In the past, I have struggled with this. I completed some assessment tools that uncovered my strengths. I reflected on things I enjoyed doing as a child, and I identified a few defining moments that clarified my values. I figured out what really matters to me. None of these things alone provided perfect answers about my passions, but each gave me insights that helped define what was right for me. Most of us feel some resistance about doing these types of activities. Your resistance might be holding you back. Stop resisting and start to allow. Allow yourself to step into your brilliance.

Suggestions

- We each have many strengths. You know when one of your strengths is at work by the way you feel when you use it. Work becomes enjoyable; it feels good. Others notice and comment on how well you do something. The more frequently you "feel good," the closer you are to your passion. When work becomes uncomfortable drudgery, you aren't where you should be, and you aren't likely using your talents. Honestly, figuring out what you excel at and what you like to do is among the most challenging work of all; thankfully, it is also the most rewarding.

- If you struggle, know that you are not alone, but also recognize that you are the only one who can do the work necessary to make a change.

In summary, you might have noticed that I have not stressed the importance of adhering to project management processes. It's not that process is unimportant, rather that adhering to process alone will not make you successful as a project manager. The entrepreneurial or "soft" skills truly make the difference.

"Process-crats" rely heavily on following standard processes, and they aren't as effective as project managers learning the art of managing people are. The most effective project managers can lead teams of people, resolve conflicts, negotiate collaboratively, and devise creative and innovative solutions to unique problems. A functional or operational manager uses these skills daily. It follows that the most effective project managers are adaptable, and they have mastered the skills required for true leadership. Little wonder they tend to be promoted out of project management roles and into more senior leadership roles.

Implement a few suggestions I've offered, and then look up for a moment. You might just see yourself becoming upwardly mobile.

Kathryn Pottruff

Kathryn is the president of Pottruff Consulting Inc., a company based in Oakville, Ontario, Canada. Pottruff Consulting helps individuals and organizations transform paralyzing complexity into dramatic results. Her clients appreciate her ability to take a complex situation and quickly devise a creative, well-conceived plan of action that leads to effective execution. That translates directly into bottom-line results!

With more than twenty years of experience as a consultant, workshop facilitator, and coach, Kathryn improves individual and organizational performance by transforming paralyzing complexity into dramatic results. She has worked with a variety of corporate, nonprofit, Aboriginal, and government organizations. Her clients are from diverse industries including pharmaceuticals, financial services, food manufacturing, insurance, consulting engineers, and more.

Over the years, Kathryn has been active in many professional associations, including the Canadian Association of Professional Speakers (CAPS), the Project Management Institute (PMI), the Canadian Association of Training and Development (CSTD) and the Company of Women Network. Kathryn believes in community, and she is a board member of the Bronte Creek Community Residents Association.

A believer in lifelong learning, Kathryn earned her MSPM at The George Washington University, a Bachelor of Business Administration from the University of Windsor, and a Bachelor of Arts from Queen's University. She is a Project Management Professional (PMP®) and a graduate of the Million Dollar Consulting College®.

Each summer, Kathryn can be found canoeing on the Restigouche River in northern New Brunswick with friends she met in a workshop years ago.

Kathryn Pottruff

Pottruff Consulting Inc.
Transforming Complexity into Dramatic Results!
2128 Emily Circle
Oakville, Ontario
Canada, L6M 0E6

E-mail: Kathryn@pottruffconsulting.com

Website: www.pottruffconsulting.com

LinkedIn: Kathryn Pottruff

Phone: 905-901-4266

The Keys to Our Success

Success—A Continuing Journey of Self-Definition, Determination, and Creating Value

By Frank Saladis

Whenever I see or hear the word success, it reminds me of a quotation I read in a book while browsing through a curio shop in Sutter's Creek, California. The quotation was attributed to Bill Cosby. He said, "I don't know what the secret to success is, but the secret to failure is to try to please everyone." I read that quotation at least twenty-five years ago, and it just seems to hang there in my mind like a constant reminder.

It happens to be very sensible and something everyone should tuck away for reference now and then. Most of us have experienced the results associated with trying to keep everyone around us happy. We have tried it at our workplace and at home with our families. Usually, we succeed in one thing, being burned ourselves while making everyone else a little unhappy and often creating conflict with people we're just trying to help.

I am unsure there is one universal set of "keys to success," but the lessons we learn in life and the lessons shared by others we admire, observe, or have great respect for can offer us some "master keys" that will help us move forward in our journey to personal and professional success.

The Master Keys

I believe there are two elements of success for project managers:

- Defining yourself in terms of your interests, your relationships, and your vision

- Value-Driven Project Management—Ensuring that intended benefits are realized and that you have created value for your clients, your team, and your organization

Having a set of master keys provides you with an opportunity to open many new doors and explore the portals you open. There is always some risk when we venture into new areas, but who, among the successful people you know, hasn't taken a few risks? Risk taking is part of life and part of the success journey we all travel.

Defining Yourself

Everyone defines success in his or her terms. We learn about success from our parents, friends, teachers, business associates, and public leaders and officials. For some, winning at all costs means success. For many, helping others succeed provides a sense of personal success. As we grow, mature, and experience what life has in store for us, we continue to revise our definition of success. Promotion from one grade to another as a child is certainly considered a success. Passing an exam, being accepted on a team, bringing a great idea to fruition, winning a game, being promoted to a higher-level position, or having a family, are all examples of success. The questions are how did you do it? What were the driving factors? Why was it important? What would you change?

We each have our reasons and methods for achieving the successes we have defined. For me, success is something I continue to pursue, and although I have had some significant accomplishments in the project management community and in my personal life, there is much more to do. The desire to do more, to explore opportunities, or the personal need to move forward or make a difference somehow and connect with more people is one of those master keys. The word is *desire*. Success comes to people who are driven (positively, of course).

When I first entered the world of management as a supervisor, I worked in an organization where people were considered secondary to the company's goals. The priority was service, and not the personnel. I bought into that philosophy and thrived in that environment.

Promotions came along as a reward for the "hard-core" approach to managing people. This approach seemed OK for a while, but it didn't provide me the sense of accomplishment I sought. I had power, but I didn't really have the respect I desired.

What caused me to begin to change my approach was a fellow manager, with the ability to establish a solid working relationship with his subordinates while still accomplishing his objectives, receiving a trophy from a subordinate. The trophy had the inscription "Coach of the Year," and the subordinate presented it as a token of sincere appreciation of listening, guiding, and believing in this person.

As the manager received the award, which was a complete surprise to him, I realized that one of those master keys was not power at all, but the ability to demonstrate to others that you are fair, compassionate, respectful, and can manage and lead.

121

I set out on a quest to change my image. It took much effort, and the attempt to change was a cause of quite a bit of skepticism from coworkers, but I continued that journey of transformation. In time, instead of people running for cover as soon as I appeared at a meeting or a work area, people came to me asking how they could become part of my team. Now, that's success!

I can assuredly proclaim that, if you are seeking success, one *master key* is your ability to work effectively with others, look for the value each person can deliver, always show respect, and to be fair, especially when making the really difficult decisions that come with the leader's responsibilities. I know many highly successful people are called leaders. They have transformed companies and made millions of dollars but, usually, at the expense of others. To me, working with people so that, if you, as the leader and decision maker, have to release people from employment for economic reasons and the person or people you let go thank you for the opportunity to have worked with you, **you** are a successful leader. There are leaders who have done this. I wish we had more like them.

Value-Driven Project Management

Project success is often described using the familiar triangle known as the Triple Constraint, which refers to being on time, within budget, and within performance specifications or scope. In some examples, quality replaces performance specifications as a success factor in the triangle. Many organizations use the Triple Constraint for measuring the value achieved because of the project, but is that really enough? Consider the following situation:

You have just completed a major project, the results have been obtained and documented, and you are preparing for the postproject review. According to the earned value analysis, you have completed the project earlier than expected and just under the planned budget reports. Your scope verification process indicates that the project team has met all deliverables as defined in the project plan.

Does this mean the project was a success and that you have met the value expectations of your client? The answer is maybe. Unquestionably, time, cost, and scope are important factors to consider, but today's business environment goes well beyond the familiar triple constraint. Achieving value is the business focus of today. What project managers deal with from project initiation to the final handoff and closure is a much wider spectrum of competing demands. These demands include:

- Quality
- Timeliness
- Delivery of the desired scope (emphasis on the word desired)
- Safety
- Aesthetics
- Employee morale
- Relationship with the client
- Cost

Determining the most important item in competing demands is the first step in understanding value. The supplier's priorities might be quite different from the client's priorities. It is essential to establish an agreed set of priorities.

Another factor to consider is what value means. If we address personal values, we might refer to things such as hard work, achievement, social justice, trust, meaningful relationships, truthfulness, courage, integrity, and family connections. From a project perspective, our values and the values established by our organizations will influence the project outcome. Success can be associated with timely completion of an activity over the work's quality. Some people practice a "good enough" approach, while others seek perfection. Some organizations intensely focus on cost, and they will sacrifice quality and scope to remain within budget. Here is another possible scenario. If you achieve all the project objectives, the client is delighted, but your team would rather quit than work on another project with you; is that success?

Let's look at an example that will help clarify what *value* means. If integrity is one of your core values and you discover a quality problem in your manufacturing process, you take the time to document the issue and inform your customer and other stakeholders quickly and honestly about the exact nature of the problem. You then determine the actions necessary to resolve the problem, the potential costs, and the time required to make the corrections. If integrity is not one of your values, you make excuses and mislead the customer. This behavior could lead to some serious problems with a client and become very costly for brand reputation and future business.

If you consider all factors, project success is a combination of agreed metrics, a set of common values clearly defined by the supplier and the client, and methodology that is practical and used. In addition, success depends on an appropriate mixture of beliefs, strong leadership, the technical skills of the team, understanding of technical requirements, being flexible to change, and listening to the specific needs of not only the client, but also the team members and other stakeholders.

Please note that although all items mentioned are important, a major success factor for any project is the project manager. The project manager is a leader, administrator, coach, mentor, entrepreneur, innovator, problem solver, team builder, and communicator. To summarize the key value characteristics of a project manager, consider the following:

Value Attitudes of the Project Manager

- **The Professional—Truthful, Ethical, Fair, Principled**. A professional approach to managing projects leads to consistency, pride, and increased brand value.

- **The Financial—Managing Cost and Efficiency**. Common-sense planning, acquiring reliable estimates, and emphasizing stewardship of organizational resources.

- **The Aesthetic—Quality and Customer Satisfaction**. The product should not only do what it was intended to do, but should also be safe to operate, be attractively packaged, and fully meet the defined and documented expectations and success criteria of the client.

- **The Social—Team Building, Relationships.** The most successful project managers understand the importance of a strong team and work toward developing synergy.

- **The Political—Influencing, Managing Conflict.** There will be conflicts during the project life cycle, and the project manager is prepared to manage the multitude of stakeholders and understand their particular influences.

- **The Leader—Sets an Example, Creates Confidence, Motivates, and Builds Loyalty.** The major value attitude in the project environment is leadership—not just the external manifestations of a leader, such as communications skills and charisma, but also the ability to create value through the participation of the team. We call this ability to create value through others "authentic leadership."

Compare these value attitudes to the projects that have been successful in your organization. You will find that your most respected and accomplished project managers continuously demonstrate these values.

The lesson learned here about project success is to emphasize continually the value of professionalism. The organizations that have matured their project management process understand that there is no substitute. The trends in project management today might be to reduce life cycle time, manage change flexibly, and to focus more on quality project execution, but clients appreciate a project manager who remains consistent, meets commitments, and creates an environment where value is defined and delivered.

Frank P. Saladis, PMP

Frank is a consultant and instructor/ facilitator in the project management profession. He is a senior trainer and consultant for the International Institute for Learning, and he has conducted many project management-training seminars domestically and internationally. He is a Project Management Professional and has been a featured presenter at the Project Management Institute® Annual Symposiums and World Congresses and many other project management events.

Frank is a graduate of the PMI Leadership Institute Masters Class, and he has held several positions in the Project Management Institute including President of the NYC Chapter, President of the Assembly of Chapter Presidents, and Chair of the Education and Training Specific Interest Group. He was editor of the internationally distributed project management newsletter for allPM. com, and he is the author of several project management books and many articles about project leadership. Frank originated International Project Management Day, and he was recognized as PMI Person of the Year for 2006.

Frank P. Saladis, PMP

President, Blue Marble Enterprizes Inc.
97 Mountain View Ave.
Staten Island, NY
USA, 10314

E-mail: fsaladis@internationalpmday.org

Phone: 718-698-0965

Managing Teams

The Keys to Our Success

CHAPTER 10

Trusting, Engaged Employees Are the Foundation of Effective Virtual Teams

By Claire Sookman and Liana Crocco

Over fifteen years spent working with and leading virtual teams, I have had the opportunity to observe, learn, and think about exactly what makes an effective, innovative, and successful virtual team that can make invaluable contributions to a company. I have discovered that much more goes into the successful management of virtual teams than you might imagine. However, I have also discovered effective strategies, tips, and techniques that lead to the development of trusting, engaged virtual teams. I want to share some of these strategies here so you, too, can master the sometimes-tricky art of virtual team creation and management.

It can be difficult to work on a virtual team. As human beings, we naturally crave the physical company of others and come to rely on facial expressions, body language, and other physical signals to learn about others and build relationships. What happens when a group of people is expected to work together without meeting face to face? It can be a daunting task because it takes a while to get used to communicating with others only virtually. However, it's not all bad news; the truth is that there are countless ways to make virtual team management easy, enjoyable, and effective.

The first step to building an effective virtual team is to build trust among virtual team members. To get a better idea of why trust is so central to the creation of effective virtual teams, I'd like to provide you with a single yet impressive statistic: A successful virtual team is 90% dependent on people and 10% dependent on technology. Yes, you read that correctly; the most cutting-edge, sophisticated technology in the world cannot rescue a team with flawed personal connections.

Take a moment to consider what that statement means. Think about the technological revolution our planet has experienced in the last ten years with the development of Smartphones and the perfection of telecommunication technology. There's Skype, IChat, Groove, Moodle, and the list goes on. What's more, we can access all this technology from our Smartphones, meaning we do not have to be chained to a desk to communicate virtually. Technology has made things much easier for us, but it is only one factor in maintaining virtual teams. The human element remains the most important factor. So, let me reiterate my most important point once more before continuing—people are more important than technology.

Now that we've established this basic ground rule, you might be wondering: "That's great, but how do we go about building these all-important personal connections among team members?" You can do this several ways. First, you must recognize the cornerstone of every successful relationship—trust. Without trust, your team simply won't function. How can team members be expected to collaborate on projects if they do not trust one another? I can't stress this point enough—trust is of the utmost importance in a virtual environment. Consider this—a study conducted by Cisco Systems tells us that it takes four times longer to build trust in a virtual environment than in a face-to-face environment.

Building Trust

You might be wondering, "How are virtual team members supposed to trust someone they've never met?" I'll admit this is a legitimate concern. It can be difficult to learn to trust people when you've never seen their face or bumped into them in the hallway interaction. Trust usually takes time and energy to develop, yet a phenomenon makes it more definitive to sidestep this issue. It's called swift trust and develops when groups of people are thrown together to complete a project, usually within a definite period.

The example I feel best illustrates this concept is a film crew. A film crew is composed of diverse people who all work toward a common goal—producing the best film possible. You have the director, actors, set designers, stuntmen, lighting crew, and countless others who all must collaborate to make a movie. These people likely haven't met before the film's production but must find a way to work together to accomplish the task set before them.

Enter swift trust. The film crew knows they do not have months to develop trust more traditionally, yet recognize the importance of trust in their working environment all the same. In situations such as these, trust seems to develop out of thin air because it is so crucial to those involved.

The same is true of virtual teams. They are fully aware that the opportunity to build trust traditionally will rarely present itself. Virtual team members might not meet face to face but must find a way to work together. As a result, trusting bonds tend to develop swiftly among team members. However, I must also emphasize that the swift trust phenomenon is temporary and only holds a team together before a more lasting form of trust can flourish. Swift trust lays the groundwork for hardworking, effective teams. So, here are a few strategies managers can implement to ensure that the swift trust phenomenon occurs on their team:

- **Align the goals and rewards of team members.** When team members believe they are working toward a common goal, they are more likely to trust one another. Emphasizing that a team will sink or swim together will boost team members' sense of dependence on one another and increase the likelihood that trusting bonds will be forged. For example, if one of your tasks involves revamping the company's website home page, pair a member of the marketing team with a writer to work on this project. In this case, the two members of the team share a goal, that is, improving the website's home page, so they are likely to trust each other because they know they must work together to get the job done. Encourage these team members to communicate by Communicator, phone, and e-mail.

- **Set clearly defined roles and responsibilities.** Stress the fact that each team member has unique talents, skills, and abilities and, therefore, a unique contribution to make to the team. The team will soon learn to respect and trust each other's unique talents and come to trust the team member who might have something they do not.

- **Create a relatively strained environment.** When demands are lax and deadlines seem looming far on the horizon, everything seems under control. As a result, motivation withers and projects suffer. Avoid this problem by creating a gentle pressure in your virtual environment. Time constraints, for example, create a sense of urgency and cause team members to rely more on one another to get the job done.

- **Treat team members professionally.**
  As a manager, you set the tone for the rest of the team. Support the notion that each team member is trustworthy. Other team members will be encouraged to follow suit. An effective way for virtual managers to encourage trust among team members during meetings is to give each team member an opportunity to share a success story or innovative idea. A virtual leader can also have team members peer mentor each other. For example, a team member strong in technology can mentor a team member not as strong. This allows each team member to let his or her expertise shine and support the other team members.

So far, we've seen the importance of building trust in a virtual environment. But what happens when we've built this trust; how else do we maintain a virtual team? Trust is only one important aspect of an effective virtual team; it is also crucial that team members constantly engage with one another and the goals of the team as a whole. This way, team members are motivated to contribute their best work. In the following section, I will explain a few strategies to maintain an engaged virtual team. Again, it's important to encourage engagement during and between virtual meetings.

Encouraging Engagement

First, what does it mean for employees to feel engaged? Briefly, it means that they should feel as though they are important members of the team with valuable contributions to make. When employees feel valued, they feel motivated to produce their best work and help the organization move forward. Engagement is especially important in a virtual setting. Think about it—it is easy for virtual employees to feel disconnected from the rest of their team.

There's no common meeting space where they can strike up a casual conversation. There's no physical meeting space they're likely to visit each day. All they have is a computer screen and a microphone. It's easy to imagine a virtual worker feeling isolated, potentially uninterested in their work, or feeling as though it's a struggle to muster the necessary motivation to get the job done.

So, how can virtual managers avoid this problem and make their employees feel part of a cohesive team? First, there are web conference applications with features that go a long way toward promoting engagement. Remember that technology alone cannot build engagement; rather, a virtual leader can use technology to encourage engagement during meetings.

During virtual meetings, leaders need to look for opportunities to verify that the people on the other end are engaged. A virtual leader should check in with everyone every three slides or every six minutes. Remember, web conference technology has tools such as chat, an electronic whiteboard, polling, and breakout rooms, which can be used to encourage collaboration and the free flow of ideas.

Another great way to build engagement is to have a virtual pizza party. You probably wonder exactly how this works. Well, arrange to have a pizza delivered to your virtual team members at lunchtime. For those team members in a different time zone not having lunch, send them a gift certificate for their local coffee shop. Gestures such as these make employees feel valued and truly part of a team. During these online parties, no work discussions are allowed; this is a time to unwind and develop personal connections.

This strategy can be used to build engagement between meetings. It is also important to build engagement during meetings as well. Check on team members by asking nonwork-related questions or by having team members share one success they had since their last meeting or a challenge with which they would like help. You might also consider posting a world map and asking virtual team members to use the pointer or arrow tool available with many web-conferencing tools to pinpoint their location. As you can imagine, these strategies require team members to become involved in pre-meeting discussions, which builds engagement.

As a virtual leader, the importance of promoting engagement was strongly demonstrated to me many years ago when I led a virtual project with team members scattered across the globe. During a virtual meeting, I walked the team through the project scope and asked them to put a Project Charter together based on the information I provided. At the end of the meeting, I asked whether there were any questions, and one, maybe two, people asked a relatively simple question, and that was it.

At that moment, I was so proud of myself because I had, or thought I had, explained the project scope perfectly. Was I ever wrong! When I received their draft of the charter, it had no resemblance to what I had said. So, what happened?

Because no one had asked any questions, I assumed everyone understood what was expected of them. I soon realized that the problem stemmed from many factors. First, I realized that, in some Asian cultures, it is considered impolite to ask too many questions. I also assumed that they understood, and during the conversation, I did not stop to recap what I had said, nor did I follow up with them after the meeting to ensure understanding.

Even though it was difficult to work through this situation, it taught me many lessons I'll share with you now. This situation taught me about the importance of engagement. After explaining the project, I directed an open-ended question to the team, along the lines of "Does everybody understand?" This might be a daunting question for some people to answer because they do not want to risk seeming foolish in front of the rest of the team if they are the only one with a question.

A better strategy would have been for me to go around the virtual meeting room and ask whether they had any questions. Or I could have followed up with each team member after the meeting to make sure he or she fully understood what was said. During the meeting, I could also have asked a few team members: "Tell me your understanding of the scope so far." Using a statement instead of a question encourages the employee to respond and, therefore, to become more engaged.

I also learned an important lesson in cultural diversity from this experience. I learned that different cultures have different values, and they might approach the workplace differently. Managers should remember this when working with team members who might be spread over the globe. Understanding the differences among cultures minimizes the chance that misunderstandings will occur among virtual team members and ensure that projects run smoothly.

Conclusion

I'd like to summarize the most important lessons I've learned as a virtual manager:

- **People are more important than technology.**
  It's important to have up-to-date technology, but this is a single factor in maintaining virtual teams. Personal connections among team members better ensure a strong, effective virtual team.

- **Trust is the cornerstone of effective virtual relationships.**
  Encourage team members to trust one another by aligning goals and rewards, setting clearly defined

roles, creating a relatively strained environment, and treating team members professionally.

- **It's important to build engagement between and during meetings.**
  Check on team members by asking nonwork-related questions or by having all team members share one success they had since their last meeting or a challenge with which they would like help. Consider also posting a world map and asking virtual team members to use the pointer or arrow tool available with many web-conferencing tools to pinpoint their location.

These strategies make virtual teams as effective as they can possibly be. If virtual team members trust one another and feel engaged with the rest of the team, they will effectively collaborate on projects and ensure the success of the team and the organization as a whole. In my experience, it is most important to focus on the human element of virtual teams. Technology will undoubtedly help your team, but people are always more important than technology. Virtual workers who trust one another and who are engaged feel like valued members of a team, and they are likely to produce excellent, innovative work that will move their company forward. Technology might be the vehicle, but the true goods are your virtual team members.

Claire Sookman

Distinguished as Canada's pioneer in virtual team building, Claire Sookman specializes in helping geographically dispersed teams reduce costs and reach their potential through training, coaching, consulting, and designing.

As the driving force behind Virtual Team Builders, Claire Sookman brings to the table more than a decade's worth of corporate and public sector training experience. In the past three years alone, she has worked with more than 4,500 managers. Specializing in virtual team building and communication strategies, Claire's personalized and focused seminars have garnered many accolades, putting her services in high demand worldwide.

Some of her clients include Boeing, Pitney Bowes, Wells Fargo, AT&T, Weyerhaeuser, CIBC, TD/Canada Trust, Siemens, Bell Canada, ComDev, Bayer, GlaxoSmithKline, Ontario Ministry of Community Safety and Correctional Services, USDA Forest Services, Hospital for Sick Kids, Canadian Public Health Service, the Canadian Forces Personnel Support Agency, the Insurance Board of Ontario, and Green Mountain Coffee Roasters.

In addition, Claire is a published author on virtual teams with five books and many articles that have been published in *CIO Magazine, Computer World, Network World,* and in *Training Report.*

Claire Sookman

Virtual Team Builders—Engaging People for Results

E-mail: csookman@virtualteambuilders.com

Website: www.virtualteambuilders.com

WordPress: http://virtualteambuilders.wordpress.com

Twitter: http://twitter.com/virtual_teams

Phone: 416-398-5160
Phone: 866.497.7749

The Keys to Our Success

CHAPTER 11

Everyone Is a VIP!

By Mark Hollingworth

The most important lesson I ever learned was to treat the people I work with as VIPs: Very Important People. Early in my career as an engineer in a unionized workplace, I spent most of my time trying to get the seven experienced technicians who reported to me to do what I wanted. If I were to do the physical work, they could file a grievance against me. Years later, I still do nothing. I just participate in conversations with others in the hope that we can ultimately make the right decisions and implement them. In general, other people implement the decisions taken. They are my VIPs. They get the job done.

What most companies state on their websites and in their annual reports is true: "Our employees are our most important asset." Most employees, however, do not feel that they are treated that way.

Being treated as a VIP is not a one-sided arrangement. In return, all employees must assume the role and responsibilities required by VIPs.

- They must adopt and build on the shared Vision of the company.

- They must have a clear, but flexible, Implementation plan.

- And they must have the People skills required to motivate their team to achieve the accepted vision and implement the plan.

Again, VIP. Easy, right?

The remainder of this chapter in the book is the story of how one young project manager learned the "VIP" lesson on his one day off work while working on a project in a strange city. I trust you will enjoy it and learn the VIP lesson along with him as his day progresses.

Everyone Is a VIP

A young project manager (PM) finally had a day off from leading a big project in a city he was visiting for the first time. He had worked night and day since he arrived and had not given much thought to how he would spend his free time.

On the morning of his minivacation, he went down for breakfast as usual in the hotel restaurant and found himself sitting next to another hotel guest. After hearing that the PM had the day free, his breakfast companion replied that he intended to spend the day at the City Art Gallery and asked the PM if he would care to join him. The idea of spending the day at the gallery was attractive, but something about the man made the PM uneasy. He was unsure he really wanted to spend his one day off with him. He politely declined the offer.

Once finished breakfast, the PM walked into the hotel lobby. A tour guide approached him and asked if he could be of service. Smiling, the PM replied that it was his day off, and he was looking to make the most of it. The guide, of course, immediately began to suggest many ideas from a long list of visitor activities. Still, the PM had no idea how he really wanted to spend his day. The guide tried his best to offer suggestions but eventually gave up, wished the PM a good day, and went off to help other hotel guests.

The PM then strolled out of the hotel and stood on the front steps, enjoying the view. Another guest exited the hotel, halted next to him, and asked if he was lost. The PM replied that he was not, but admitted that he had not yet decided what he wanted to do that day. The woman responded by inviting him to join her on a trip to the shopping mall. Having nothing better to do, the PM joined her. The woman walked rather quickly, taking shortcuts through car parks and down back streets. She obviously knew where she was going and how to get there. Arriving at the shopping mall, the PM realized that it held little interest for him and so proffered a friendly good-bye to the woman. He felt a little frustrated that he had wasted the early part of his day.

He looked around and saw crowds of people rushing in different directions. They all seemed very busy, pressed for time, and eager to get where they were going. The PM observed them with some amusement, but also with a little envy. They all seemed so focused on their goal. He had no idea where he was going—and even felt a bit lost.

Amid all the activity, he caught the eye of a woman desperately trying to catch the attention of a passerby.

"Do you know how to get to the Houses of Parliament?" she asked. "I'm afraid I don't," he answered.

Undaunted, she replied, "I have been trying to get there since I arrived in town, but still don't know how to get there. Would you like to come with me?" "Um, no thanks," he replied, taking his leave of the woman.

At a loss for a better idea, the PM decided to go on his own to the City Art Gallery. The PM approached a passerby and asked her if she knew the way to the gallery. The woman had only good things to tell him about the gallery's wonderful design and the great works currently on display. Enthused, he asked again how to get there. She began to give him vague directions before admitting rather sheepishly that she had never been there but had heard great things about it just recently from her friends. Somewhat disappointed, the PM thanked her and walked away. He asked two more people if they could help, but they, too, could not provide directions.

Finally, he entered a nearby hotel to seek directions. The concierge was chatting to two other hotel guests. The PM overheard their conversation, and he was intrigued. He immediately had a good feeling about the guests and recognized that they were in a similar position to his. Indeed, it seemed that the PM had quite a few interests in common with these fellow tourists, including art.

The PM introduced himself, and it soon became clear that the three visitors would have much more enjoyment spending the day together than if they went their separate ways. With the concierge's help, they made a list of activities they all wanted to do and developed a clear plan of how to get to each different attraction. By the time they assembled their itinerary and left the hotel, all three were excited about the great day they had ahead of them.

During the morning, they explored and enjoyed some fascinating places on their list and a few others that they discovered and entered as they were passing. Unfortunately, while stopping for lunch at a restaurant, they learned that the City Art Gallery, which they had planned to visit in the afternoon, was closed for the day because of unforeseen repair work. While all three were disappointed, they soon realized that they would now have more time to visit other interesting tourist sites in the area. After they finished their lunch, the three companions jumped on a bus to take them to a historic chateau on the edge of town.

By the end of the day, the PM and his two new companions had spent a wonderful day together and had formed a friendship that would undoubtedly lead to more meet-ups in the future. The PM returned to his hotel exhausted but feeling very satisfied with his one-day minivacation.

As usual, the PM sat down later that evening to write in his journal and reflect on the lessons of the day. He recognized that every person he had met that day had something to teach him.

He quickly realized that he had trusted some people he had met, but not others. For example, he knew that he had made the right decision not to spend the day with the man he had met over breakfast. The required level of trust had simply not been there. He also understood that the tour guide could not help him because he himself had not given enough prior thought to what he was really looking for. The PM ultimately realized that the reasons he had gone along with his two new friends and had enjoyed the day so much could be summed as follows:

- First, they built a shared vision of where they wanted to go. By taking the time to converse with one another before heading out, they developed a common goal and understanding of what they all wanted to get out of the day. They identified and discussed the many attractions and activities the city had to offer before coming to a collective decision about what could best be achieved in a single day. It was an ambitious objective but motivational and feasible. By the time they set off, no one had imposed his or her vision on the others.

- Second, they developed an implementation plan focused on getting them where they had decided to go, yet they remained flexible enough to modify it as the day unfolded. The PM realized that during their initial conversation, they had established a series of guidelines based on their shared personal objectives, interests, and personalities. These guiding principles directed them during the day, particularly when their original plan was affected by unpredictable outside influences. This had proved essential after they discovered that one of their underlying assumptions—that the City Art Gallery would be open—had proved false.

- Third, the three companions formed a great team for the day by being trustworthy, being open to the others' ideas, and through having open dialogue. Each knew that any one of them could simply walk away if they could not agree on the decisions made or failed to achieve a "win-win" for all concerned. As a result, a true feeling of fellowship and camaraderie developed.

The PM realized that much of the day's success depended on their building a common, realistic, yet exciting, *Vision*; identifying a clear but flexible plan of *Implementation*; and creating a team of three great *People* who communicated well together on all issues.

"*VIP*," said the PM, smiling to himself.

He realized that, on his free day in the city, he had succeeded in identifying the three key skills that form the essence of a great project management team and distinguish a great team leader. He had met people with great vision, but no plan. He had seen scores of busy, preoccupied people, but with whom he shared no common vision. He had also met a man over breakfast with a good vision and a plan but who lacked that essential "P" factor. The PM knew that all three were imperative if he wanted to become a great project team leader. Did he himself have these three VIP skills and attributes?

Already, even during his short career, how many times had he witnessed team leaders and members talk a wonderful vision of what they would do—but who, at the end of the day, had no real implementation plan and produced very little? The cliché about the need to "walk the talk" existed for a reason.

How many times had he also come across teams where each person worked hard but without any shared vision of where the team as a whole was going? These teams always underperformed or delivered undesired results in the end. He also knew that there were many teams out there with great visions, goals, and objectives, and a detailed plan for what needed to be done to achieve them, but because of hidden agendas, internal team conflicts, poor communication, and low motivation and trust, they would never achieve anything of note.

Perhaps, he considered, by building his VIP abilities and by being a role model for such behavior, he could help his team members at work feel and behave like VIPs too. What a team he would have if all the members behaved like VIPs, and in return, they were treated as VIPs! Every corporate website and annual report says that employees are the company's most valuable asset, but how many employees feel they really are VIPs? The young PM was determined to be the first to make this slogan a reality for his team members.

In his journal, he summarized his new learning. He wrote, "Everyone is a VIP."

"*Vision*" means creating a clear shared vision of who we are and where we will be two to three years from now.

"*Implementation*" means having a rigorous but flexible plan, road map, project portfolio, PMO, and so on, that show how to get to the vision.

"*People*" means assembling and leading the team who buy in to the vision and implementation plan and who make a full commitment to making sure the necessary work is done.

149

Excited, he headed to the hotel restaurant for dinner. He could hardly wait to get back to his project the following day.

Ten Questions to Consider about Your VIP Skills

Vision

1. Do you have a clear vision of who you are and where you are going in life?

2. Do you explore all possible options before making a decision, or do you choose the first solution that comes to mind?

3. Have you built a shared vision with your team members about where the team is going?

Implementation

4. Do you have a clear plan and key performance indicators in place for your team, which establish the steps required to achieve your vision?

5. Does your implementation plan assume you know how the future will unfold? Will it still be workable if uncertainties and major changes occur?

6. Do you put tight rules in place that your team members must follow, or do you establish guiding principles that provide them with the latitude to deal with problems appropriately, as they arise?

People

7. Do people feel free to express their true thoughts and feelings in the meetings you chair, or do they censor themselves and say what they think you want to hear?

8. Does your reward system reward the behaviors you seek to promote within your team?

9. Are you really worthy of being trusted?

VIP

10. Would your colleagues and team members at work agree that you really treat them as VIPs?

Mark Hollingworth, MEng, EMBA

Mark is the president of 5i Strategic Affairs and an expert in strategic leadership. He has worked as a management consultant, workshop leader and facilitator, educator, and ontological coach to the profit and not-for-profit sectors in Montreal, Canada, for more than twenty-five years.

Mark's mission is to help "people, companies, and society prepare for the future." He is the strategic planning process leader, guide, trusted supporter, and devil's advocate to ensure that individuals and organizations make the best decisions possible—and that those decisions result in concrete action and change.

In the past five years, he has worked with companies such as Cogeco, Hydro-Quebec, Genfoot, Imperial Tobacco, Kruger Inc, Linamar, Metso Minerals, RioTintoAlcan, RSM Richter Chamberland and Standard Life and has worked abroad with clients in Abu Dhabi, Bhutan, Costa Rica, Ghana, Guatemala, Kenya, Nigeria, Saudi Arabia, Spain, the US, Vietnam, Zimbabwe, and so on.

Mark is a faculty lecturer in the School of Continuing Studies and a program leader in the Executive Institute at the Desautels Faculty of Management at McGill University where he leads the "Strategic Planning & Execution" and "Creativity & Innovation to Create Value" seminars.

He is author of the book *Growing People, Growing Companies: Achieving Individual and Organizational Success in the Knowledge Economy* (2003) and, since then, has also had several articles published in the *Ivey Business Journal.*

Mark Hollingworth, MEng, EMBA

President
5i Strategic Affairs Inc.

Faculty member
The Executive Institute, Desautels Faculty of Management, McGill University

E-mail: Mark.Hollingworth@5istrategicaffairs.com
E-mail: Mark.hollingworth@mail.mcgill.ca

Website: www.5istrategicaffairs.com

LinkedIn: mark-hollingworth

Phone: 450-510-5024

Business Outcomes

The Keys to Our Success

CHAPTER 12

Become "Project Business-Savvy"

By Gary Heerkens

As you progress through your project management career, most tips and guidance you will receive are likely to pertain to how you can manage your project more efficiently and effectively. That's good—every organization you work for will certainly expect you to direct the execution of your projects with great skill.

But for each project you are assigned to manage, there is one critical question you should ask—and keep asking —from the moment you are assigned to manage the project throughout its life: Why is my organization doing this project? Your ability to answer this question competently and correctly—and to use that answer to direct the course of your project properly—will be greatly enhanced if you develop the ability to think (and act) like a businessperson. Therefore, my #1 critical tip and key to long-term success as a project manager is to become project business-savvy.

The immense importance and value that comes in answering the why question stems from recognizing that it really doesn't matter how well a project is executed if it is the wrong project—a project your organization should not even pursue.

The reason for this is simple: Fundamentally, organizations pursue projects to achieve one (or more) of these objectives:

- To fulfill the organization's strategic intent or strategic mission

- To optimize the organization's operations and/or processes

- To make or save money

These are often called high-level business drivers. And even though project management is often viewed as a set of tools to manage execution logistics, the reality is that decisions made throughout the life of a project can influence how much that project's business drivers are satisfied. Project managers who recognize this—and who understand how to make project decisions that will optimize the fulfillment of business drivers—will be highly valued for many years.

In this chapter, I'd like to discuss three key developmental aspects of becoming "project business-savvy":

1. What does it mean to be project business-savvy?

2. Why should you want to become project business-savvy?

3. How can you develop and implement a project business-savvy approach?

What Does It Mean to Be Project Business-Savvy?

Perhaps the best way to conceptualize how the term business-savvy pertains to a project manager is to use your imagination. Imagine that you're in charge of a particular project. Instead of leading a technical, deliverable-focused project (which is the way most project managers are conditioned to think today), imagine that you are proposing, starting, then operating a business venture that requires a significant financial investment. Now, consider the kinds of things you would need to be concerned with about that business venture. If you stretch your imagination, you will realize that your concerns would extend far beyond what they are today and might include

- The profitability of the venture

- How the venture will be financed

- Marketing and promoting the venture's outputs

- Operating the venture after startup

- Legal and ethical considerations

- The general economy and its effects on the venture

- The existence and resultant impact of business risks

Now, if we translate this back to a project context, it's fair to say that you would be expected to perform and think more like an executive than a project manager. For some, this can be a daunting task. But those who live up to this expectation can also expect many new avenues of professional growth to become available to them.

From the perspective of managing projects on a day-to-day basis, being project business-savvy refers to project managers who understand how and why to incorporate a business perspective into their decision-making. They exploit every opportunity to optimize positive strategic, operational, and financial outcomes throughout the project investment life cycle. To do this, business-savvy project managers must be competent in:

- **Business Knowledge.** Knowing which business concepts and principles have relevance to projects and to project management. For example, a business-savvy project manager knows what the Weighted Average Cost of Capital is, and how it applies to the financial justification of projects.

- **Business Skills.** Skill refers to the ability to perform. The business-savvy project manager doesn't just know about things related to business and projects—he or she is able to actually do things. As an example, this means he or she would be fully competent in leading the effort to prepare a clear, concise, and comprehensive project Business Case document, if asked.

- **Business Acumen.** This means knowing when and how to apply business knowledge and skill to a given project situation. For example, if a risk surfaced during his or her project, the business-savvy project manager would analyze the risk and base his or her response on factors such as financial or economic impact, strategic impact, process optimization, value maximization, and cost vs. benefit.

Being project business-savvy might also open new avenues of organizational involvement—and more professional development opportunities for you! The knowledge, skill, and acumen described above can be applied to many activities that typify the intersection of the world of business and the world of projects and project management. Among the more notable activities are the following:

- Analysis of customer business needs
- Project opportunity identification

- Strategic planning and strategic management
- Project evaluation and selection
- Project financial analysis
- Business case preparation
- Return on Investment (ROI) studies
- Cost vs. benefit analysis
- Portfolio management
- Tracking and measuring actual benefits realization

Promoting the notion that project management practitioners can be valuable contributors to these kinds of critical organizational functions is a positive movement. And it promises significant gains for the organization, for us all as project management professionals, and for you as a practicing project manager, which leads me to my next key point.

Why Should You Want to Become Project Business-Savvy?

For me, this question was answered in the early stage of my career in managing projects for the Eastman Kodak Company. My MBA studies made me begin to think much more like a businessperson. I started to become much more aware of organizations that simply pursued projects without giving much specific consideration to the question of "Why?" This led to some tough questions, which triggered much deeper thinking by my clients about the concept of project benefits. The result was the development of a list of projects having a much greater "bang for the buck."

Through this experience (and many others), I recognized a simple reality—project managers who understand the connection between the world of projects and the world of business are extremely valuable—and will certainly be highly regarded in the future. They also have the power to transform the project management landscape, and many project management experts agree that a world full of business-savvy project managers represents the future of the profession.

This has now evolved into a kind of "movement," and the movement has two key drivers. First, more and more companies are coming to realize that projects are financial investments—key agents in the quest for strategic, operational, and economic success—and should be treated as such. Second, more and more people who care about the project management profession (such as I!) are seeking ways to elevate its stature and increase the respect for project management practitioners everywhere. And although it's important to recognize this movement, let's not forget that being business-savvy will undoubtedly enhance your prospects for career advancement!

But some still question whether it is the project manager's job to understand the business aspects of his or her project and to be involved in business-related activities. Those who pose this question point out that engaging in processes and practices relating to the business side of project management—such as business case preparation and portfolio management activities—is not within the typical project manager's formal job description. For many organizations, this is true—but the number of organizations that think this way is dwindling. They recognize that project managers who lack an understanding of the business drivers of their projects are likely to make generally poorer decisions throughout the project's life.

But no matter what your organization's position is on whose job it is, there remain many excellent reasons you should take the initiative to learn proactively about (and participate in) the business and strategic aspects of the projects you're assigned to manage. Here are just a few:

- **It can be a key enabler of your career advancement.**
  Project managers who demonstrate concern for and understanding of business issues are likely to be viewed very favorably when the time comes to hand out promotions. In today's competitive environment, being business-savvy is a strong differentiator when comparing competencies.

- **You are likely to become involved in a broad range of interesting activities.**
  Practicing business-based project management might include anything from helping prepare project business cases to participating in strategic planning sessions (as described above). Many of today's intelligent, able, and ambitious project managers would welcome these kinds of opportunities.

- **Your project decision-making ability will improve dramatically.**
  Basing project decisions only on technical or functional considerations means that not all the critical inputs required to make the best possible decision are considered. Project managers who do not understand their project's business aspects are destined to make suboptimal decisions from time-to-time.

- **Respect for you—and your role as a project manager—will increase.**
  Besides giving you an immediate career boost, practicing business-based project management will help us all. It will demonstrate that project managers can contribute much, much more than many are being permitted to contribute today.

- **You can make a valuable contribution to our organization's success.**
  Organizations that do not recognize the critical connect points between their projects and their

business are likely to waste considerable money and human resources. Project managers who practice a business-savvy approach can help organizations recognize these connections and reduce the waste.

How Can You Develop and Implement a Project Business-Savvy Approach?

Ultimately, the widespread adoption of a business-savvy approach toward projects is an organization-wide consideration. Ensuring that practices such as project financial analysis and business case preparation become regular parts of doing project business must be driven and supported by senior management. And inviting the project management community to participate on "front-end" activities such as strategic planning, operational planning, and portfolio management is largely at the discretion of senior management.

However, as an individual project management practitioner, you can do some things on your own—right now—to demonstrate your business savvy and advance the cause of business-savvy project management in your organization. Here are just a few:

- Verify that your project is clearly linked to (and driven by) one or more key organizational strategies, business objectives, or operational goals. Make sure your management knows that you are aware of these links and that you will integrate them with your project decision-making. If the links are not made clear to you, don't be afraid to ask what they are.

- Strive to understand deeply the business benefits your project yields. For example, does your project: Advance long-term strategy? Increase revenue? Lower operating costs? Reduce headcount? Increase process efficiency? Leverage this knowledge to make appropriate and high-quality project decisions.

- Ensure that the person identified as the project sponsor is a business sponsor. A business sponsor is the person who has direct responsibility for the achievement of the project's stated business benefits (as described above), not just technical success.

- Cultivate a good working relationship with your organization's "financial person" (or group). This is particularly helpful if he or she is a regular participant in activities such as project financial analysis; ask him or her to teach and/or help you understand the financial analysis process in your organization, if one exists.

- Ask a senior manager with an impressive "business mind" if he or she would be willing to be your mentor. Tell him or her that you are striving to bring a business-savvy perspective to your management of projects. This is one of the best ways to gain significant knowledge, while offering you the opportunity to interact with a member of senior management.

As you implement a project business-savvy approach, you may also be helping your organization develop its project business savvy because a surprising reality exists in today's project environment: many organizations are either not aware of—or not knowledgeable of—many techniques that compose business-savvy project management. One explanation for this is that those organizations tend to think of projects as "part of doing business"—in other words, they tend to view projects as a subset of ongoing operations.

As project management practitioners, we know that this is untrue; every project has a distinct start and a distinct finish. A much more critical point, though, is that these organizations do not necessarily think of projects as discrete investment opportunities. Opening your management's eyes to this reality might be an important learning opportunity for them—and a way to demonstrate that you are a deep thinker!

Becoming project business-savvy will likely open excellent opportunities for you to lead your organization into the future, thereby demonstrating tremendous incremental value for you personally, and for project management in general. For example, if your organization does not currently prepare project business cases, consider drafting a business case for the next project you are assigned to manage (I'd suggest that you make it a brief, but effective, example). Present your document to the appropriate members of management, pointing out that preparing project business cases can offer assurances that your organization is investing limited project resources wisely—and avoiding projects that are not good investments.

In Summary...

Rapidly advancing to the forefront in the world of project management today is the realization that projects are key agents in nearly every organization's quest to achieve positive business results. And many organizations realize that—within reason—it doesn't matter how good the schedule is, or how well the people in the project are managed interpersonally, if those aspects of project management are applied to initiatives that represent "poor business."

Becoming project business-savvy represents a unique, timely, and valuable opportunity for you to set yourself apart in today's project management environment. And even if business-oriented skills and competencies are not specifically identified in the formal job description for project manager in your organization, the expectation might still exist. And even if this expectation doesn't yet exist, project managers who demonstrate the initiative to practice business-based project management are likely to be greatly appreciated and highly valued as they carry their organization into the world of "next level project management." In short, they will be viewed as superior project managers. You will be one of them if you become project business-savvy!

Gary R. Heerkens, PMP, CPM, MPM, CPC, CBM, CIPA, PEng, MBA

Gary is the president of Management Solutions Group, a firm that provides progressive and effective project management training and consulting solutions. Before establishing Management Solutions Group, he managed projects for Eastman Kodak for more than twenty years, and he was staff assistant to Kodak's director of project management, developing and delivering project management training, developing project process methodology, and consulting within Kodak.

Mr. Heerkens is a longstanding seminar provider for Project Management Institute's SeminarsWorld series and is the project-management education provider for the Purchasing Management Association of Canada. He has been invited to speak at twelve of the last thirteen PMI Global Congresses, including encore requests in 2007 and 2008.

Mr. Heerkens is a contributing editor to PMI's *PM Network* magazine, writing on "the business of projects." He authored McGraw-Hill's *Briefcase Book on Project*

Management, still a leading seller, and *The Business-Savvy Project Manager*, published in 2006.

Mr. Heerkens is an active member of several professional societies, including PMI, IAPPM, APBM, AAPM, and IPMA, where he has received several professional certifications; he has been honored as a Fellow by IAPPM and by AAPM. He was president of the PMI Rochester Chapter from 1998–2001 and was reelected president in 2005–2006. He's a licensed professional engineer in New York and holds an MBA from RIT.

Gary R. Heerkens, PMP, CPM, MPM, CPC, CBM, CIPA, PEng, MBA

Management Solutions Group, Inc.
208 Kim Lane
Rochester, NY
USA, 14626

E-mail: msginc@frontiernet.net

Website: www.4msginc.com

LinkedIn: www.linkedin.com/pub/gary-r-heerkens-mpm-cpc-pmp-cpm-cbm-cipa-peng/5/1a5/904/

Phone: 585-820-3660

CHAPTER 13

Business Outcome Management

By Ken Robertson

At the completion of most projects, the project manager and the team celebrate getting to the finish line. Most of these project managers see the finish line as completing all the project tasks and being ready to move on to the next project. What is often overlooked is whether the expected project business outcome was achieved. It is gratifying that the team is satisfied with completion, but did the project deliver to the business' expectations? The key lesson I've learned is that the business leaders who sponsor projects do not look favorably on project managers who ignore the business outcome produced. This can affect a project manager's ability to be chosen to run the next high-profile business project.

This chapter reviews business outcome management—clearly defining expected business outcomes, tracking progress toward these outcomes, reaching the project "finish line," and realizing/sustaining the business outcomes. Following these concepts enhances the actual and perceived

167

successes of project managers and project sponsors in directly affecting the business they support, which can be a significant step forward in enhancing your credibility as the project manager.

Concepts

There are many proven methodologies for project management, all geared around structuring your project such that the project manager can effectively initiate, plan, execute, control, and close projects. Today's project manager is trained in how to "grind out" the project—work the plan and finish within the budget, schedule, and quality expectations. Unfortunately, many project managers do not clearly define the business outcomes expected from the project.

The Project Management Institute (PMI) defines nine knowledge areas in its *PMBoK Guide (Guide to the Project Management Book of Knowledge)* essential to managing a project successfully. What I've learned is that there probably should be a tenth knowledge area that focuses on business outcome management. As you can see in the following diagram, this new knowledge area is reflected in the project's time frame—it should be one of the first things a project manager considers and should be the primary element in assessing the project's relative success.

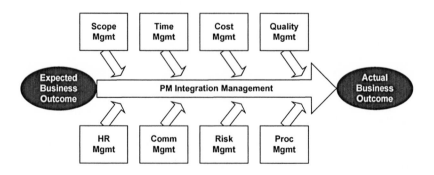

Defining Expected Business Outcomes

During a project's initiation stage, the project manager
works with the project sponsor, project team, and end
customer to define clearly what the project will deliver.
In most cases, this involves ensuring that technical
specifications are clearly documented and agreed. Although
this is a critical step, it should be augmented with a broader
discussion of what the customer expects to achieve from a
business perspective.

The expected business outcomes are not a repeat of the
technical specifications and not just about completing the
project. They are true business results such as

- Enhancing customer satisfaction and/or loyalty

- Reducing costs by...

- Increasing revenues by...

- Increasing market share by...

- Achieving a return on investment of...

- Achieving a positive net present value

If a business case has been prepared for the project, some
specific business outcomes have likely been identified. You
might find these in the financial analysis, the business
strategy and/or in the executive summary describing what
the project will deliver to the business. It is now critical
for the project manager to ensure everyone on the project
clearly understands what is expected.

The project manager, on your own, cannot develop
these expected business outcomes. You need to engage the
business to define these. You need to discuss and document
them in terms clearly understood by the business. Each of
these outcomes needs to be clear, concise, and measurable—
something for the project manager and the project sponsor
to track.

One approach I like to take with customers is to help them build a "mental video" of what success looks like to them and to the business as a whole. You can work with your customers to paint the picture of what they want to accomplish, how people will feel about it, and how this translates into an enhanced position for the business. This process helps both the project team and the business to understand clearly what the project is expecting to deliver. This mental video should be used extensively throughout the project and should be integrated into your communications plan, risk management plan, and change management process.

On a recent workflow management project, our team developed a mental video of what the experience would be like when a customer arrived at the business counter for service. We designed the processes to meet the target experience, shared this video with customers in our communications to them, and integrated it into everything we did on the project. When any change was proposed, we replayed the video to see how the change enhanced our customer experience—if it didn't, we denied the change. This simple concept can have a powerful impact.

This story-telling approach can be a powerful tool in your project. It helps all your stakeholders understand what is delivered and sets clear business expectations for everyone. The process also helps your project sponsor thoroughly understand why he or she is doing the project and what the business expects from the project. In many cases, project sponsors consider their role in projects as requiring limited time, merely to sign things off. This recommended story-telling approach engages them more, and it should be fully incorporated in the project sponsor's roles and responsibilities for the project.

Tracking Business Outcomes

Once the expected business outcomes have been established, the project manager is now in a position to track the project against these expected outcomes. One key element of project management is progressive elaboration— we learn more as the project progresses, allowing us to learn and solidify our plans, costs, and expectations. The same applies to expected business outcomes. What we defined as the expected business outcomes at the project's beginning undoubtedly change as the project progresses. It is, therefore, essential for the project manager to continue to work with the project sponsor discussing, reviewing, and revising the expected business outcomes throughout the project.

If you define expected business outcomes at the project's beginning and then forget about them, you will likely find that the final business outcomes will not align with the original expectations. More important, the business might determine at the project's end that the business case is no longer valid for the project and that investment has produced little or no return. They will wonder why you, as a project manager, didn't bring this forward.

Project managers should thoroughly review the impact on the expected business outcomes as part of the change management process. When a change is proposed to a project, project managers should ask themselves:

- Does this change enhance our ability to deliver the expected business outcomes?

- Does it detract from our ability to deliver?

- If yes to either question, then by how much?

Although these questions might seem simple, they are typically difficult to answer. The project manager needs to engage the project sponsor and key business resources thoroughly to review the business outcomes. You need to run through the mental video with the adjustments for the proposed change. Revisit the financial analysis for the project if financial expectations are major drivers. Be prepared to do a full business assessment of the change— not just a technical analysis of the impact to the project delivery, timeline, scope, and budget.

Project managers should review the project's ability to achieve the expected business outcomes monthly in addition to doing this as part of the change management process. This effort should be treated in the same way project managers should review their project risks to determine if there is any change and, therefore, any adjustments that might be needed.

Remember, we constantly learn during the project, and this knowledge needs to be applied to the analysis done in the beginning to keep the business expectations aligned with the project's ability to deliver. This might require a change to the project scope to ensure the business expectations can be delivered—this might be downsizing the scope to match declining business returns better or upsizing the scope to exploit ascending business returns more thoroughly.

As project managers, we tend to drive through our projects to the "bitter end," based on a belief that success is defined by just completing the project tasks. With this new emphasis on business outcome management, we need to think of success in achieving the expected business outcomes. If we cannot acceptably achieve these, then our project is not a success. Project managers would be best served to work closely with their project sponsor and senior management potentially to close down or downsize projects that will not deliver business value. This can be a tough step for project managers but one that truly separates the exceptional ones from the rest.

Project Completion

As you reach your project's full completion, you are ready to turn the "product" of your project over to the business. If you defined the expected business outcomes at the beginning, tracked them along the way, adjusted the business expectations and/or the project, you should be in an excellent position to deliver to the business what they expected. The mental video created by the business will come to fruition, and all stakeholders should be satisfied.

For many projects, the realization of the business outcomes will happen significantly after the project is completed. For example, if the project is to develop a new product for the marketplace, then the success of the initiative will only be determined once the sales force has the product available to sell. In essence, the project objectives are to place the business in the best position possible to achieve the expected business outcomes, but in the final analysis, it is up to the business to deliver.

The following lifecycle diagram shows the overlap between the project and business value processes. The business needs to start its own business project to drive the expected business outcomes. Business executives and line managers whose objective will be to use, sustain, and exploit the business benefits presented by the project will

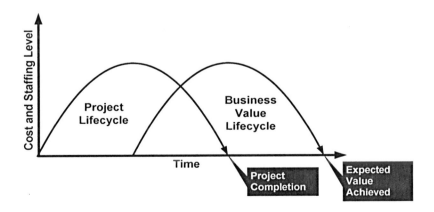

run this project. Project sponsors must assign resources with specific time frames, objectives, and budgets to drive toward the realization of these business benefits.

Part of the normal project management closing process should include a full audit and accounting for the delivery of expected business outcomes. This includes reviewing the original expectations, the revisions made along the way, and the results produced by the project. As with other aspects of the project, a full lessons-learned process should be applied. This should include assessing the reality of the business expectations, the engagement level of the business, the business case presented, and the likelihood of fully realizing these benefits. The lessons learned should be shared throughout the business to ensure these ideas are integrated into future projects.

I have performed postproject reviews on projects where the business has been fully engaged along the way and the expectations clearly tracked throughout the project. In these projects, the business typically takes the project's product and operationalizes it to the point where most of the expected business outcomes can be achieved. In these situations, proactive work by the project manager results in a smooth transition to the business. When the business is not fully engaged in the project and business outcomes are not "top of mind" during the project, the transition tends to be very cumbersome, and the business' ability to operationalize and achieve the expected business outcomes is minimal, at best.

Summary

As project managers, we need to remember that it's not just about the project management process; it's about the business result, which is, after all, why the project was started in the first place! Sometimes, we focus too much on our project's technical aspects because that is where we think the real complexity is. But for the people investing in your project, it is more about the expected business outcomes—how your project will enhance the customer's business.

The process described in this chapter is not that onerous. Although it takes some time, it is extremely valuable in engaging the business in your project and in ensuring you and your team do not lose sight of the true purpose for the project. It should also be stressed that many project managers see this as just an extra overhead process that is not needed. Like the risk management process, it might not have an immediate return, but it will be beneficial at some point in the project.

In thirty years of managing projects, I've yet to see a project that didn't have a risk pop up or have some change in the business expectations. So, as the saying goes, "You can pay me now... or pay me later," but you will pay.

Project managers who focus on business outcomes find that their project sponsors are more responsive, the recipients of the project are more engaged, and the project's final success is significantly enhanced. They also find that senior management is more confident in their project portfolio when they know proactive decisions are made based on expected business outcomes, not just on a technical team's desire to grind a project to completion.

Project managers also find that this process helps their technical resources better understand the business world and the impact of what they do. It lifts them from a technical-centric view to hopefully one more business-focused. Project managers themselves find that their value to the organization increases when they incorporate these ideas. It is a simple concept that is a "win-win" for everyone; you just need to take the time and the initiative to make it happen.

Ken Robertson, MBA, PMP

Ken is an experienced project manager, writer, and lecturer. His company, KLR Consulting, specializes in project management for business, technology, and real estate development projects. Ken has been a project, program, and portfolio manager for more than thirty years. His experience includes working with a broad range of private sector companies and with all levels of government.

Ken focuses on helping his clients get the most out of their project investments. His business outcome approach resonates well with the senior executives he works with regularly.

Ken is the coauthor of the book, *Project Management Methodology: A Practical Guide for the New Millennium*, and author of *Work Transformation: Planning and Implementing the New Workplace*. Ken holds an MBA from Simon Fraser University. He is a certified Project Management Professional (PMP) through the Project Management Institute (PMI) and an Information Technology Certified Professional (ITCP) through the Canadian Information Processing Society (CIPS). Ken is a regular writer and presenter on the topics of project management and strategic business planning.

Ken Robertson

KLR Consulting Inc.
Burnaby, B.C.
Canada

E-mail: ken.robertson@KLR.com

Website: www.KLR.com

LinkedIn: Ken Robertson

The Keys to Our Success

Change

The Keys to Our Success

CHAPTER 14

Leadership Impacts on Change

By William S. Bates

All projects deliver change at the organizational and personal levels. The various stakeholders can perceive the impact of that change as positive or disruptive.

Change can stir strong emotions among employees and management. It can make or break careers. Largely, success or failure of the delivered change depends on the leadership provided for the project at both the project manager and project sponsor levels. Too many project managers tend to focus on changes to delivering the project's scope and overlook the most important change involving projects—the change in an organization caused by implementing projects.

To ensure project success, a key role of a project manager is to manage project stakeholders' organizational change expectations. To ensure the focus on organization impacts caused by various changes implemented by a project, the project manager must also provide the necessary team leadership and indoctrination to provide team focus on these change impacts.

. It is up to the project manager to bring the project team leaders together and make them recognize themselves as the core of the team who always realize their responsibility to the organization the project will affect. This is especially true for the situation of a dispersed team working from several separated locations to achieve a common result, which is quite often the case today on larger projects. I have encountered this situation in two mission critical projects with a large number of stakeholders, where in both cases, stakeholders included the companies' customers, suppliers, and their management and employees.

The first was an airline's reservation and departure control system. In this case, the project team of more than 270 people was spread across several major cities in Canada and London, England. Four large companies and several smaller contractors provided the team. The airline company itself had three large teams from marketing, departure control, and information technology who provided the key stakeholders' view within the airline.

These teams were very focused on the new system's impacts on their functional groups. The project delivered five months early, a million dollars under budget with a successful and expected product. Although many factors involved caused it all to come together successfully, the key was the project director's leadership approach in ensuring a strong voice for the airline's chief impacted functional groups. With a friendly, but firm approach, he made sure that constant coordination and communications among the various subteams and their members were always foremost.

The second was a project rescue on a ten-million-dollar project that had existed for almost three years with no end in sight. The project concerned a new application systems platform that could cause the company to fail in customer expectations if not delivered in the near term. The company was a French company operationally headquartered in the US with parts of the project team in Canada, France, and the US.

The frustrated company management contracted a project management team to bring the project back on track. The contracted project management team found twenty-one small teams each working semi-independently without a central project manager and little planning in place. Leadership was required to consolidate the large number of small teams, develop a coordinated and tractable project plan, win over the key team leads, and open clear communications with all internal stakeholders. That undertaking of leadership saw the initial release of the product delivered within seven months of the contracted project management team arriving.

In both cases, leadership and management of stakeholder expectations played major roles in achieving organizational change success. So, what is leadership? I define leadership "as the ability of a person to influence and cause other people to move in concert to achieve some goal or objective." Leadership combines a balance of understanding people, goals, common sense, and judgment used to cause others to follow. Of importance in leading project-driven change is to remember that a major goal of the project should be that the stakeholders accept the change being brought about by the project and the change absorbed with minimum disruption, if possible.

Too many people believe position implies leadership. Nothing could be further from the truth. Being appointed a project manager does not imply that person appointed is an able leader. Certainly, it is expected by their management and their team that they will be a leader, especially in those common situations where the project manager does not have much, if any, real authority to direct. Team members who are not managers might provide good leadership.

Obviously, leadership is applied in as many environments as there are human endeavor—family, community, political, military, government, commercial, academic, and so on. Very few people are born with true leadership ability. Even those who have more natural leader ability traits than others still need to gain experience and grow to develop their true leadership abilities further. How do we then gain our leaders?

Leadership ability is gained through experience and learning across a period. For successful leaders, the leadership learning process never stops. The learning process involves formal training in dealing with people as well as the observations of other leaders and managers for concrete examples of both good and bad leadership and management practices. Leadership development should also be progressive, commencing with the leadership of small projects before undertaking large projects.

- The leadership learning process, besides on-the-job, should include formal training in such subjects as

  ° Team building

  ° Conflict management

  ° Psychology of leadership

- Ideally, the leadership experience process should evolve from leadership of small work package teams through ever larger leader assignments to project director of major projects. All these leader assignments might not necessarily be projects, but also include operational manager positions. Major mistakes are sometimes made by organizations who assign a senior technical person to be project manager of a major technical project with a large team. This usually leads to less than desired performance by the overall project team primarily based on the lack of leadership skills in the project manager. Even worse is their heavy devotion to the technical aspects of the project deliverable as opposed to the needs of the stakeholders who must live with the change.

The paramount goal is the striving for leadership for the greater good of our stakeholders at all levels. Whether leadership is superior, good, or bad depends on the character base of the leader in question. That character base begins its development during childhood from family, school, religion, and interaction with his or her peers. When seeking managers of any kind, in addition to their experience in their field, evidence of their character is also necessary. How do we determine character?

The required character factors in leaders are:

- The integrity not to compromise your beliefs as well as the trust of those involved. These attributes are based on sound moral principles, uprightness, honesty, and sincerity.

- Integrity involves telling a straight story, be it good or not so good news. Never lie, not even little white ones. Become known for telling it like it is.

- Keep your stakeholders informed of the situation and direction despite it being good or bad news. Surprises usually help downgrade your credibility.

- Loyalty to their people and clients as well as the nation, community, and organization. Loyalty has to work in all directions. Project managers must be loyal to their team, their management above, to their suppliers and contractors, and their stakeholders, and they should be able to expect loyalty in return.

- In the Airline Reservation System project, the project director took special care to show loyalty to all teams despite the home organization. This caused loyalty back to the overall team, which in turn led to the project's success.

- Leaders show respect to the team, clients, suppliers, culture, environment, organization, and country.

- In the Applications System Platform project, the contractor project director made extra efforts to show his respect and trust of the various teams working on the project by recognizing the mixture of cultures and languages involved. He did this by meeting with his team leads on their home ground (Canada, France, and the US), listening to their input, and integrating their inputs where possible.

- Success in any venture, project or otherwise, is based on trust among the people involved. Project managers must protect their character by always being ethical and respectful in their dealings.

- In working with the dispersed teams on the Applications Systems Platform project, the project director made a point of visiting each site in Canada and France, getting to know the team leaders and members, and helping with their planning, which helped gain the teams' trust.

- Leaders need to be fair and evenhanded in their dealings with their team, suppliers, clients, and superiors.

- Especially with the project team, this means providing evenhanded treatment of individuals or subteams and not showing favorites or partisanship. With the twenty-one subteams in the Applications Systems Platform project, the project director was always mindful of this need and always ensured an evenhanded approach by him and his project office team.

- Leaders must have the courage always to "do the right thing," to take assessed business risks as needed, and to have the conviction of their beliefs.

- In an off-track and mismanaged stock exchange project, a project management consultant was called in. His primary role was to analyze the project background and history. The resulting report caused the project to be cancelled after thirty million dollars had been spent and cost three executives their jobs.

- In an oil refinery company information systems project, the project manager announced in an executive management meeting concerning the project that the project was stopped. The explanation was made that the executives needed to find time to provide requirements guidance critical to proceeding. The executive involvement was then provided.

- Finally, leaders need to be ethical in their dealings with everyone.

- In a large federal government, a male project planner was accused of improper behavior toward several female members of that planner group. The project director immediately started a formal and open investigation of the accusation and arrived at a resolution that satisfied all parties involved while keeping the issue in the open for the team.

It is critical how a project manager interacts with all project stakeholders. Those interactions can make dramatic differences in how those various stakeholders respond to changes in their environment. At all levels of stakeholders, there must be recognition that they have differing needs and expectations. The project manager's job is to balance these differing needs and expectations.

When there are inconsistencies to work out, the best approach must satisfy as many stakeholders as possible and provide explanation and communications to all. Thus, an understanding of the project's priorities and needs plus balancing of the project stakeholders' needs is required.

One approach to gaining stakeholder support is to explore alternative approaches with the different groups of stakeholders, aiming for consensus of approach.

The stakeholder groups usually of most importance are:

- The organization's executives and senior managers whose operations, directly or indirectly, are affected by the changes brought to play. Here, the project manager's leadership becomes critical as he or she interacts with these managers. The project manager's credibility and the seniors' level of trust of the project manager could have much impact in resolving differing expectations and acceptance by this senior management group.

- Lesson learned—In a large company with a monopoly market, in a large project to develop and install project management methodology and process, the project had a steering committee of senior managers. Failure to understand the expectations and concerns of these managers led to a less than spectacularly successful implementation. The underlying cause was the CFO's belief that the project should have been under his sponsorship as opposed to that of the EVP of Operations.

- The project team is also a critical group of stakeholders. A team that is handled fairly, treated with respect, and shown obvious loyalty is far more likely to become very team-focused and drive to deliver a quality product on time. A driven team whose project manager does not protect them and treat them fairly will usually not be as focused and dedicated.

- The Airline Reservation System project's director went to great lengths to know and understand his teams and members. He travelled to meet with

the various teams and occasionally participated in
afterhours get-togethers with the working teams.
They knew who he was. This team from several
companies repaid his devotion to them with an early,
under-budget finish.

- Impacted working-level organizational staff can
make or break the acceptance of the change caused
by the project.

- In the late 1960s, a major military personnel
system was developed that significantly changed
the way personnel data was entered and updated in
the system. Data entry moved from typed reports
converted to punched cards to optical character
typed reports read directly by the system.

The system near failure was found to be based on a high
percentage of errors introduced into the system. It was
discovered that clerks responsible for data entry had been
ignored in training preparation for the system. During the
parallel implementation of the system, these untrained
clerks were required to double enter all data in the old way
and the new way. Only immediate training of the clerks and
forced turn-off of the old system saved the system.

This problem caused more than a year's delay in the
system's final acceptance. The object lesson here is to
remember the lowest-level employees whose interaction
with the system must be considered for consultation and
training.

Other stakeholders who can have influence, depending
on the project, are customers, the public, politicians, the
competition, other projects competing for resources, and
other organizations that could be affected. In some way, a
project manager should consider consultation with these
various groups. This could be accomplished through focus
groups, surveys, and meetings with other project managers
and functional managers to resolve resource conflicts.

In projects crossing organizational boundaries, several departments of a government or multiple companies involved in delivering a project, the project manager must establish relationships with the senior management of each entity at the necessary levels to gain their support and participation. Project managers' credibility and success in accomplishing projects depends on their accomplishing this exercise.

Remember; a project that delivers on time, in budget, and with the technical scope as defined, but which has disrupted the organization and has substantial resistance is a failure. How project managers exercise their leadership skills and abilities greatly affects the degree to which the various stakeholders accept the organizational changes caused by a project. Management of stakeholder expectation and requirements is one of the most important factors in accepting the changes caused by the project. Frequent and factual communications are essential for success.

William S. (Bill) Bates,
Executive Consultant, Trainer, Speaker

Bill is currently the CEO of a project management services company. He is the retired president of Bates Project Management Inc., a private company with a business history of thirty-four years, specializing in project management services. Bill is also a retired officer of the United States Marine Corps. He has more than fifty-five years of leadership and management experience.

He has more than thirty-eight years of experience in applying the project management process across various types of projects in industry, government, and nonprofit environments. In addition to authoring and designing several project management methodologies and software systems, Bill has developed project-management training programs and delivered more than eight hundred seminars internationally. His portfolio of project management achievements includes successes in engineering, social services, aerospace, high technology, facilities and accommodations, information systems, manufacturing, government, oil and gas, mining, construction, and many more.

He holds his BS from the University of Maryland and MA in the Technology of Management from the American University.

Bill is active in professional organizations, and he has served on the Board of Directors of the Ottawa Valley PMI Chapter. He has been a member of the Project Management Council of the Conference Board of Canada, and he frequently delivers presentations for local PMI Chapters' Project World conferences. He has served on the Research Advisory Committee of the Project Management Institute

(PMI). He has authored articles in such publications as *Datamation, Computing Canada,* and *The Government Computer* magazine.

William S. (Bill) Bates

Autonomous_ID Canada Inc.
ECBD Consulting Inc.
3181 Riverside Drive
Ottawa, Ontario
Canada, K1V 8N8

E-mail: billbates@autonomousid.com

Website: www.autonomousid.com

Phone: 613-680-6462

Delivery

The Keys to Our Success

CHAPTER 15

Scalability and Common Sense: Key Ingredients for Success

By Paul Bergman

How many of us have been frustrated on a project because we had to go through the same procurement process for a $1,000 item as a $100,000 item? Or perhaps we have been expected to create a risk management plan for a complex project in the same time as for a simple project. Organizations often require project management tools to be applied in the same way across a broad range of projects or applied only to projects valued at greater than a specific dollar figure or project duration. All other projects are then left to the tried and true FBTSOYP (Fly by the Seat of Your Pants) style of project management. These practices have the potential to turn project management into an exercise in frustration.

According to the *PMBOK Guide*®, "A project is a temporary endeavor undertaken to create a unique product, service, or result"—this definition is simple to grasp, straightforward, and not too controversial. It is also worthwhile to think about what the definition does not say. Notice that the definition of a project makes no mention of duration. For example, the definition does not say, "A project is a temporary endeavor and at least 'x' months long." There is also no mention of cost or the number of people involved. The fact is that projects come in all shapes and sizes, from the multiyear billion-dollar projects that involve thousands of people to short projects that involve few people and cost no money (such as volunteers giving their time). To remain relevant, our project management tools and level of rigor need to be scaled to support the variety of projects we encounter.

The Importance of Results-Oriented Project Management

In our zeal to ensure that project management tools are applied, we sometimes forget why we are applying them in the first place. The following example illustrates this problem.

A contractor, whose experience involved small projects, was asked to prepare a Gantt chart (a commonly used representation of a project schedule) for the construction of an office building. The client wanted to ensure that the contractor would deliver a successful project with few surprises. The contractor created a Gantt chart that looked very professional and met all the client's target milestones. However, when creating the Gantt chart, the contractor did not break the project down to a manageable level of detail, consider task dependencies, consider resource availability, or allow any contingency. The contractor had not done any formal planning for scope, risk, or quality management. The Gantt chart provided little, if any, evidence that the project would be properly managed and would be successful.

In the above example, the client asked their contractor for evidence of project management but made the common mistake of assuming that a project schedule was synonymous with evidence of the application of proper project management. The contractor, on the other hand, used tools of project management but presented only an illusion of a properly planned project. The contractor likely did not realize that although less formal application of project management tools might have sufficed for his smaller projects, for a project of this size, the tools would have to be scaled and applied more rigorously if the project was to be successful.

The proper scaling of project management tools requires us to go beyond their superficial applications. We need to start with clarity—what is the value expected from the application of the tools? We can then ensure that the value is supported by the level of project management applied. So, how do we start?

Back to Basics

Before a meaningful discussion on project management scalability and common sense can happen, organizations must come to terms with how formal project management tools provide value. Some benefits of applying formal project management tools and techniques, as opposed to FBTSOYP project management, include

- Deliver projects to strict deadlines
- Control project costs within narrow profit margins
- Reduce exposure to liability resulting from unidentified risk or a poorly defined project scope
- Obtain stakeholder input and buy-in
- Coordinate the activities of multiple resources
- Provide a framework for accountability and reduce abuse (for example, inflated hours or costs)

- Achieve quality assurance on the project
- Optimize resource utilization
- Reduce disruption to key business processes
- Document the process so that
    - Similar projects undertaken in the future have a template to work from
    - The project is compliant with industry or regulatory standards (for example, a pharmaceutical company involved in a new drug development project)

Once we come to terms with *the value* of project management, we can begin to scale our use of project management tools and techniques to our project requirements.

If our project is not time-sensitive, for example, we might benefit from de-emphasizing the schedule in favor of greater controls on cost, quality, scope, or resources. That doesn't mean that schedule targets should not be set, but simply that deadlines for such a project should not be the primary consideration when making decisions, and the schedule need not be planned in great detail.

If, on the other hand, time is of the essence and the sooner a project is completed the more profitable it will be (as is often the case with product launches or scheduled maintenance shutdowns), more detail should be used when developing the schedule. We should also adjust the change control procedures so changes that can save time—even if they incur extra cost—can be evaluated and approved quickly.

Apply Common Sense

The successful application of project management is all about balance—applying the tools and level of rigor that provide the control and certainty that will bring value to the project. When we forget balance, project management transforms from a powerful tool to a wasteful bureaucratic process.

How do we achieve the right balance? We start by trusting our common sense when applying project management tools. For example, common sense should tell us that for a new product development project in which customer focus groups will be used to determine key product features, we will not be able to define the entire project down to a "manageable" level of detail at the start. Instead, we draw on experience to define the project at a high level; break down deliverables in the near future to a manageable level; then allow the needed time, resources, and money to refine the plan further as more information becomes available.

This common-sense approach is a form of progressive elaboration and represents the rolling wave approach. The rolling wave approach recognizes that, for many projects, only after a project begins to take shape do we have sufficient information to plan the details of what ultimately needs to be done. This approach is not the same as taking it day-by-day. A rolling wave approach requires us to recognize that some project details become available only as the project unfolds. However, no aspect of the project should be implemented before having been defined to a manageable level of detail.

Encourage Scalability of Tools

What else can be done to achieve the right balance of project rigor? We can convert our project management forms into templates. What's the difference? Forms are generally rigid. They are created to collect specific information (think of tax forms as an example). Templates are designed to provide a starting point—to make our job easier by not having us start from a blank sheet. Efficient use of a template requires an understanding of purpose and intent and encourages customization based on the specific application and context. A template, therefore, implies flexibility based on need.

Take, for example, a Project Charter. A Project Charter is a valuable project management tool developed during the initial stages of a project. Key inputs to the development of a Project Charter include a Statement of Work (a narrative description of products or services to be delivered by the project) and a Business Case. The Project Charter might also include:

- Clarification of high-level scope or deliverables and of the gray areas around what is included and what is not included

- High-level requirements for achieving project success

- High-level information to help decide whether a project is realistic and worth pursuing

The Project Charter is not, however, a detailed project management plan. On the contrary, it is designed to ensure we are on the right track prior to detailed planning (which can be very costly and time-consuming depending on the project). Properly constructed, a Project Charter saves time and money and can make the difference between a project that gets off on the right foot and runs efficiently and a project that experiences constant flux as we run around in circles trying to satisfy multiple stakeholders whose ideas of the very purpose of the project are not aligned.

Because the Project Charter is a core tool of project management, many organizations create predefined forms for their Project Charters—often requiring detailed project information. They then mandate that the entire Project Charter form be filled in before the project can receive approval to proceed to the detailed planning phase. Large or small, simple or complex, every project is expected to begin with the completion of the same Project Charter form.

Problems develop when the standard Project Charter does not allow important information or when the information demanded by the standard form does not apply to a given project. Problems also arise when the standard form demands a level of detail or accuracy simply not available at such an early stage of the project. Project managers learn that if they don't complete all fields on the form, the project will be held up until the entire form has been completed. On the other hand, once the form has been completed, experience tells them that the Project Charter will almost certainly be approved, and the project will proceed. When this occurs, the information contained in the Project Charter becomes less important than the act of completing the Charter. Rather than a critical tool of project management that helps filter less worthy projects and ensures that key project stakeholders are on the same page, the Project Charter is reduced to a "paper exercise."

By replacing project forms with templates, we return the focus to the tool's purpose and the information required to make it useful. A project template might contain some required fields, but when using the template, project managers are encouraged by organizations to scale the information found on the template to what is relevant to the project at hand. It might even be useful to offer different starting templates for different types of projects. For example, a Project Charter template for new product development projects might include clear go/no-go decision points, whereas a Project Charter template designed for a project that is one of several interdependent projects might require a section on related projects.

A similar approach should be applied to the application of all project management tools, including risk management, scope management, quality management, contract and procurement management, and the degree of project monitoring and control we place on our projects throughout their lifecycle.

Conclusion

To be successful project managers, we need to expand our focus beyond the application of project management tools. We should have in mind the purpose and value of the tools, the target audience, and the success factors most important to the project being managed. If common sense tells us that we will not achieve the desired results given the way we are applying the tools, we need to adjust our application of the tools—scaling the tools to provide the value and control suitable for the project.

Use of scalable tools and common sense when applying them are essential to supporting project management and project success. When we get that right, positive results follow.

Paul Bergman

In 1986, as a recent engineering graduate working for an aerospace manufacturing company, Paul was asked to manage his first significant project. Not knowing anything at the time about "formal" project management, Paul turned to the IT manager for advice in selecting a software tool that could help manage the project. Paul soon learned that there was more to project management than simply using project management software.

Although Paul's first opportunity to manage a project was not as successful as it could have (or should have) been, he pulled through with enough lessons learned to be given more project management opportunities. Within a few years, his experience grew to include the management of many manufacturing, IT, environmental, and HR projects. In 1992, Paul obtained his PMP® designation.

In 1995, Paul founded World Class Productivity Inc. (WCP) and developed the first version of what was to become the company's flagship three-day Project Management Essentials workshop—believed to be the longest continuously running project management course in Canada, with more than 8,000 participants to-date.

Today, WCP has grown to include several of Canada's top project management trainers and consultants. Our courses have been delivered in-house to more than 100 of Canada's leading organizations, and we are featured in three university-level project management certificates. Paul's signature style for course development and delivery, in which realistic scenarios are used to challenge participants' application and understanding of the principles, is evident in all his courses.

Paul Bergman, PMP

President and Senior Consultant

World Class Productivity (WCP) Inc.
201 Franklin Ave.
Thornhill, Ontario
Canada, L4J 7L4

E-mail: pbergman@wcpconsulting.com

Website: www.wcpconsulting.com

LinkedIn: paulbergmanpmp

Phone: 905-660-7184 ext. 101
Phone: 800-214-8096 ext. 101

CHAPTER 16

Moving from Waterfall to Agile

By Patrick van Abbema

Many of today's project management and business analyst professionals find themselves leading, managing, and analyzing on Agile development teams only to find that many tools and techniques applied when using a traditional project management approach no longer work as effectively or at all. To do more than survive in this iterative development environment, I have learned that project managers and business analysts must use additional project management and business analysis tools and techniques to lead their teams effectively and deliver their projects.

There are many challenges to shifting to an Agile framework. Key points to keep in mind are:

- It is difficult to change. Not everyone will adapt.
- Communication is critical to success. You must engage everyone.
- The key roles must have people of influence.

- It requires discipline and consistency.

- Be prepared to show tangible benefits early and often.

This chapter will explore how your projects can transition easily and successfully to an effective Agile environment.

Agile is an incremental, iterative framework for project management where requirements and solutions evolve through collaboration among self-organizing cross-functional teams. This disciplined process involves:

- A leadership philosophy that encourages teamwork, self-organization, and accountability

- A set of engineering best practices intended to allow rapid delivery of high-quality software

- A business approach that aligns development with customer needs and company goals

Your role in an Agile project will look much different as you form and coach a self-directed team, facilitate continuous collaboration with your clients, and manage and deliver business value to your clients early and regularly throughout the project.

Comparison Between Agile and Waterfall

There are many debates about Waterfall as opposed to Agile development methodologies, but before deciding which is appropriate, it is important to understand both of these approaches.

Waterfall

A traditional linear and sequential approach to software design and systems development, each waterfall stage is assigned to a separate team to ensure greater project and deadline control. A traditional linear approach means a stage-by-stage approach for product building, for example:

1. The project team first analyzes, then determines and prioritizes business requirements and needs.

2. The design-phase business requirements are translated into solutions, and decisions are made on how to move forward.

3. Processes are defined and online layouts built; code implementation occurs.

4. Work evolves into a fully tested solution for implementation and testing for evaluation by the end-user.

5. The final stage involves evaluation and maintenance, ensuring everything runs smoothly.

What I have observed with using a Waterfall approach is that there is a high cost to change, and it focuses on change control (for example, change against a plan based on many assumptions) as opposed to change management (for example, evolving the plan as the team better understands the solution that will meet the requirements)

Agile

Agile is a low overhead method that emphasizes values and principles. Working in cycles (for example, every two weeks), project priorities are reevaluated at the end of each cycle. Four principles that constitute Agile methods are:

1. People and interactions over processes and tools

2. Working software over comprehensive documentation

3. Customer collaboration over contract negotiation

4. Responding to change over plan follow-throughs

To summarize the difference between the two, the Waterfall method stands for predictability, while Agile focuses on adaptability. Agile methods are good at reducing overheads, such as spending time rationalizing, justifying, documenting, and meetings, keeping them as low as is possible. Agile methods can benefit small teams with constantly changing requirements.

Agile methodology means cutting down the big picture into smaller puzzle size bits, fitting them together when the time is right, for example, design, coding, and testing bits. By managing in smaller chunks, you harness changing requirements and communicate these changes to key players more effectively

I have observed that when a stage is completed in the Waterfall method, there is no going back planned, because most software designed and implemented under the Waterfall method is difficult to change according to time, the plan, and user needs. The problem can only be fixed by going back and potentially designing an entirely new system, a very costly and inefficient method. An Agile framework, designed to cope and adapt to new ideas from the outset, adapts to change, as at the end of each stage, and allows changes to be made easily. With Agile, changes can be made if necessary without getting the entire project rewritten. This approach not only reduces overheads; it also helps manage risks and costs.

Another Agile method advantage is having a tangible product at the end of each tested stage, which ensures bugs are caught and eliminated in the development cycle, and the product is tested again after the first bug elimination. This is impossible for the Waterfall method, because the product is tested only at the end, which means any bugs found result in the entire program potentially having to be rewritten.

Agile's modular nature means using better-suited, object-oriented designs and programs, which means you always have a working model for timely release, even when it does not always entirely match customer specifications.

Whereas, there is only one principal release in the Waterfall method and any problems or delays mean highly dissatisfied customers.

Agile methods allow specification changes by end-user requirements, spelling customer satisfaction. As already mentioned, this is impossible when the Waterfall method is used, because any changes made potentially mean the project (or many parts of it) has to be started all over.

Addressing Challenges

From working on both types of projects, I've realized that two inherently different styles and cultures present many challenges to the project team. Of course, working with two styles (and cultures) of project management presents many challenges to the project team. Key points to consider when adapting to Agile are:

Terminology

A key challenge in transitioning from a Waterfall process to an Agile one is using common terms that everyone understands. A ScrumMaster must coach team members in this transition, both in using traditional Waterfall terms to the client and in minding the gap between the Agile world and their counterparts in the Waterfall world. With this transition in place, the project team can have more freedom to adopt their processes and take the necessary steps described in Agile.

Consistent Rhythm of Work

The project team's schedule will be in a sprint cycle constantly—whether they are specifically working on a release candidate. Therefore, the sprints (for example, two-week increments) cover the sizing work, production support activity, and qualify work. Because of this, the team tries not to have the sprints start and stop on just one stage of the project.

Sprints are essentially consistent units of time (also known as a timebox) used throughout the project. These consistent timeboxes allow the team to plan activities and forecast the effort for those activities. With a focus on immediate and tangible goals, they help the team quickly identify issues that need to be addressed. Similarly, it is useful not to have the sprint start or end on a particular phase because sprints are consistent units of time, despite the project phase.

Managing Incidents

The project team can struggle with handling incidents in the sprint. Some team members can be dedicated to handle upper-level support, but members across the team must deal with most incidents. Although all this work is usually not fully planned, a general level of support-related work should be expected.

This work is dealt with in two areas: analysis and resolution. A guideline that helps move the team forward with integrating production support with the Scrum model is to set the priority of any incident to the highest possible value, thus initial analysis has the highest priority over anything else on a project member's plate or task list.

Additional Tips for Migrating to Agile

Here are three additional tips when moving from Waterfall to Agile:

- Ensure that the product owner and ScrumMaster are people of influence. It is not easy to get everyone to adapt to Agile, and these two people must have the necessary influence to succeed with the transition to Agile.

- Accountability and Transparency—Collaborate with the team to identify clearly who is responsible for what.

- Early and Often—Build a sustainable rhythm with the team to deliver tangible product items or services in each sprint. The team shows progress by delivering early and often.

Realizing Benefits

Some benefits my projects have realized by shifting to an Agile approach are an improvement in work quality, faster implementation, greater customer satisfaction, and better processes. Project teams can achieve other benefits as well, including some not usually thought of as Agile benefits:

Managing Activities

A key principle of Scrum is its recognition that during a project the customers can change their minds about what they want and need and that unpredicted challenges cannot be easily dealt with in a traditional predictive or planned manner.

The daily Scrum—Each day during the sprint, a project status meeting occurs. This is called *a daily scrum, or the daily standup.* This meeting has specific guidelines:

- The meeting should happen at the same location and same time every day.

- The meeting length is set (timeboxed) to fifteen minutes.

- All are welcome, but normally, only the core roles speak.

- During the meeting, each team member answers three questions:
    - What have you completed since yesterday?
    - What are you planning to complete today?
    - Any there any impediments that prevent you from completing your work?

A key benefit of adapting Scrum is that it enables the team to fit large items into releases. This might seem too much a challenge, because limiting sprints to two weeks eliminates the ability to handle large pieces of functionality. On any project, large features are difficult to fit into the prioritized list for a release.

Scrum forces the team to regard these big features incrementally, breaking them into analysis or research, and then incremental implementation. The team fits the analysis or research into sprints not committed to the development phase, providing them with the opportunity to finish the up-front tasks and making it easier to fit these activities into the release priorities.

Related to this is the benefit of gaining the ability and discipline to deconstruct large tasks; sprints force the development team to think in smaller tasks. With smaller tasks, the team can plan release content better and reallocate resources if tasks slip. Large tasks tend to require a broader range of skills and expertise, but when broken into smaller tasks, many of these could be assigned to individual team members with specific skills, enabling greater flexibility in assigning resources. After several releases and several sprint-planning sessions, projects will see this "small-task" thinking become ingrained on the entire team.

This approach can be used for nonfunctional items as well, such as code refactoring and automated integration testing. Convincing the client that such work is important will be challenging, but the team can break the refactoring work into small pieces, build a gradual plan for the changes, and then fit the incremental changes into sprints and releases, the result of which will increase the speed and flexibility of product solution development.

Important: The team needs the support of the product owner and other executive champions to gain the agreement to move forward with this approach.

Improving processes

The Scrum process provides a structure for incremental process improvement with which teams are usually challenged. The sprint retrospective meetings provide useful feedback, and process changes are added to the next sprint. Keep in mind; it can take several sprints to refine the processes to the point of success. The team will commit to the process, work through the issues, and build processes that will work well for the team.

Change Management

The team can become better able to deal with change when it occurs. For example, during a release, the team can begin with a clear understanding of the priorities, and then the client can come forward with a change request or new feature they want to add. Typically, such disruptions could affect a team, but they use the principles of Scrum to talk about the options (for example, reprioritizing the product backlog). Because the project team can make these decisions, they can adapt better and can accept the new work with clear success metrics.

Improving Team Health

Agile Scrum can positively affect the team's morale and overall health. Because the project team can improve their processes, they are committed to the processes, and they put energy and enthusiasm into their work. Creativity can then be applied to process improvements and the team's development efforts. Updating processes can greatly improve the project cycle, reducing the anxiety when a milestone is due, and reducing the number and severity of problems found. The release process then becomes more a steady, well-planned effort. Work hours are then more predictable, as the team hit key milestone dates, which improves morale and reduces tension among the members of the team working on the project.

Conclusion

Thus, Waterfall's defined stages allow thorough planning, especially for logical design, implementation, and deployment. Agile methodology is a good choice for projects dealing with uncertainty and requirements that seem to change constantly.

If you are interested in implementing a more Agile mindset to your projects, I encourage you to read the following articles and books:

- *Agile and Iterative Development: A Manager's Guide* by Craig Larman

- *Agile Estimating and Planning* by Mike Cohn

- *Agile Project Management with Scrum* by Ken Schwaber

- *Agile Retrospectives* by Esther Derby and Diana Larsen

- *Agile Software Development Ecosystems* by Jim Highsmith

- *Agile Software Development with Scrum* by Ken Schwaber and Mike Beedle
- *Scrum and the Enterprise* by Ken Schwaber
- *User Stories Applied for Agile Software Development* by Mike Cohn
- Many weekly articles at www.scrumalliance.org
- PMI's Agile Certified Practitioner (ACP) website http://www.pmi.org/Certification/New-PMI-Agile-Certification.aspx

Patrick van Abbema

Patrick has more than twenty years of progressive accomplishments in competitive digital media, Web collaboration, and Enterprise/ SaaS software markets. Patrick is the senior project advisor and chief business analyst for AltNexus Corp. and provides consulting expertise on enterprise service strategies for various public- and private-sector clients across the US, Canada, and Europe. He is responsible for analyzing the business needs of his clients, and he is a liaison among stakeholders to elicit, analyze, communicate, and validate requirements for changes to business processes, policies, and information systems. He recommends solutions that enable organizations to achieve their business goals.

Patrick is a Certified Business Analysis Professional (CBAP®), Certified Scrum Professional (CSP), and Project

Management Professional (PMP®). Patrick has a Bachelor of Arts and a Master's Certificate in Business Analysis from the Sprott School of Business, Carleton University. Patrick's experience and credentials allow him to bring the unique skills and knowledge areas required for both project manager and business analyst disciplines to the table, ensuring that the product will be on time, on budget, and most important, on requirements.

Patrick van Abbema, PMP, CBAP, CSP

Senior Project Advisor/Chief Business Analyst
AltNexus Corporation

Website: http://www.AltNexus.com

LinkedIn: http://www.linkedin.com/in/pvanabbema

Phone: 613-316-3244
Fax: 866-314-7916

Stakeholders

The Keys to Our Success

CHAPTER 17

Get in the Head of your Sponsor

By Kevin Aguanno

Over the years working at IBM, I performed dozens of project audits for both healthy and troubled projects. These assessments of project health and risk look at many factors, one of which is the way the project manager (PM) interacts with the project sponsor and other key stakeholders. By examining how the PM approaches these key relationships, I can get an early indication of possible issues where stakeholders would not be satisfied with project outcomes. In addition, I can assess the maturity of the PM, as more experienced PMs take a very different approach to these relationships than junior managers do. In my experience, properly managing sponsor and stakeholder relationships is one of the most important competencies a PM can demonstrate.

In this chapter, I show you how the most successful PMs approach these key business relationships.

Start by Understanding the Business Case

You can plan your project many ways that respect your schedule, budget, scope, and/or quality constraints. Successful PMs don't just take the easiest approach or the one they used last time; rather, they deliberately choose a project delivery approach that will maximize the delivery of business value for the project at hand.

Of course, this means that the PM must first understand the business case, something few junior PMs attempt but nearly all the most successful PMs focus on early in their projects. By business case, I am not only talking about the minimum scope, key milestone dates, and budgetary constraints, but also factors such as expected benefits, when those benefits should be achieved, the impact on other interrelated initiatives, how the project aligns with broader organizational strategies, and more.

To gather this information, the PM has to work more closely with the project sponsor in a way that could push many junior PMs out of their comfort zone. The PM needs to speak to the sponsor in business terms, such as increasing revenues, reducing wait times, improved operational efficiencies, reduced risk, and so on, not domain-specific technical terms that the delivery team might use, but which mean little to a nontechnical business person.

I have told my sponsors that I can plan the project many ways, all which will meet their constraints, but the approaches might each affect the achievement of the business case results differently. I ask the sponsors to help me choose an approach that will maximize the benefits they need from the projects.

Don't Forget the Politics

Although there is the "official" business case—the one shown to the stakeholders in the funding organization—there is also an "unofficial" business case. This second business case is rarely written and is often never spoken of aloud. It includes the secondary benefits the project can realize. For example, how (or to whom) the project reports progress might help the sponsor gain support for the project within the organization; how the project is delivered might help the sponsor meet his or her workplace performance objectives for the year, earning the sponsor a bonus or a promotion; or the timing of the delivery of a project's work products might support other initiatives.

What we often talk about in this area is **office politics**. I want to know what will help make my sponsor and other key stakeholders successful in their jobs. Sometimes, what matters might not even be what criteria are on my project sponsor's annual performance review, rather what his or her boss' performance review criteria are. If you can help make the sponsor and other stakeholders successful in other areas through how you deliver your project, then you will gain their full support.

To get this information, the PM has to establish **trust** quickly by showing that he or she has the sponsor's best interests in mind. Explain fully the various project approaches and both the direct and indirect benefits of each approach. By working with the sponsor collaboratively to plan the project approach, you will get the sponsor's support and trust.

Analyze the Stakeholders' Needs

Stakeholders—also called "interested parties" in some circles—include all organizations and individuals with an interest in the success (or failure) of your project. Stakeholders might be directly involved in the project or might be involved behind the scenes, with their influence felt through the words and actions of third parties. There might even be stakeholders who have been forgotten and who are just waiting for the project to get into its later stages before making their presence known.

The entire list of project stakeholders includes both the project sponsor and the project team members; these latter two groups are called "**internal** stakeholders" or, sometimes, "core stakeholders." Internal stakeholders have a personal stake in the project. Perhaps the perceived success of the project will affect their yearly performance reviews.

Those who are not **internal** stakeholders are, by definition, external stakeholders. An **external** stakeholder might have an interest in the project outcome, but does not have direct daily involvement in the project.

The distinction between internal and external stakeholders can best be highlighted using an old joke:

A pig and a chicken decide to open an all-day breakfast restaurant together. The chicken suggests that it be called "Eggs and Ham," but the pig adamantly insists that it be called "Ham and Eggs." When the chicken asks why the pig insists on being first in the name, the pig says, "Because you are only **involved**; I am **committed**."

The joke makes the point that "pigs" (internal stakeholders) "have skin in the game"—have a direct, vested, personal interest in the projects' success. "Chickens," although still important to the project's success, don't have the same involvement or commitment; in essence, they enter the picture, drop an egg (introduce a requirement), and then walk away, expecting the pigs to address the new requirement.

A technique that helps identify stakeholders and their relationships to the project is called **stakeholder mapping**. In this technique, a PM prepares a traditional organization chart of the people involved (directly or indirectly) in the project. Then, in the boxes for each person, the PM identifies the role of each person on the project (if they are directly involved), the level of support for the project (positive or negative) demonstrated by the person, the level of contact the PM has had with each person (no, little, moderate, or much contact), whether each person is an internal or external stakeholder, and who has political power or influence on the project. Using this information, the PM can understand better the office politics surrounding the project, decide which stakeholders should receive more attention from the project team, plan activities to engage some stakeholders better, and prioritize (possibly competing) stakeholder requirements.

Stakeholder mapping might also help identify missed stakeholders; for example, a PM or business analyst who takes the time to think through what the early postproject period should look like might realize that no one from the operational support organization has yet been engaged by the project team. In this case, the PM would add a box to the stakeholder map to represent the missing role, and then, the PM would begin to meet with the sponsor or other stakeholders to identify who would likely be the person or group assigned to that role. Once a contact is found, the PM can then begin to discuss the project with the new stakeholder and begin to set expectations and capture requirements.

Finally, the stakeholder map should be reviewed and updated regularly to ensure that changes in the project organization or in stakeholder representatives are reflected in the document. New representatives might mean new requirements, new visions, and new expectations. You must be diligent in keeping current with the changes in the project's stakeholder network.

Through structured analysis of stakeholders, their needs, and expectations, a PM or business analyst can ensure that the project has the best chance of receiving support from the broader community of affected parties, giving the project an improved chance of success.

Manage Expectations

The adage goes: Put ten project stakeholders in a room, and you'll find twelve different views of what the project will accomplish. Not surprisingly, this often turns out true. Even though the project might have a clearly documented project scope and a well-defined project plan, there might still be very different interpretations of what the project deliverables will look like and what goals the project is trying to accomplish.

PMs need to listen keenly for any signs that the sponsor or other stakeholders might understand the expected project outcomes differently. Missed expectations lead to unhappy stakeholders. Unhappy stakeholders can lead to a deemed project failure. Missing the expectations of even one stakeholder might significantly challenge a project, as the stakeholder might now become hostile and start to work against the project team. Identifying any misunderstandings or varying interpretations and resolving any resulting issues is a valuable activity a PM or business analyst can do to help keep all stakeholder expectations in line with the planned project delivery.

Be Seen as a Partner or Trusted Advisor

By talking in business (not technical) terms and planning for the delivery of the maximum business benefits, including indirect benefits to the sponsor, other stakeholders, and other projects, the PM will be seen as a trusted partner to the sponsor, rather than as a solution provider. In an era when businesses seek to reduce budgets in all cost centers, it is smart for PMs to change the business' perception of the profession from one of optimizing the delivery of a project to minimize the spending of funds to one where the PM is seen as the vehicle for achieving business value from projects. I have seen many successful PMs make this shift; afterward, they are seen as trusted advisors—even leaders—and their careers advance to higher levels as a result.

Conclusion

Although you can be successful as a PM just delivering the scope within the time and budget you have been given, the most experienced PMs—those with the greatest levels of career success—go beyond these simple constraints. By understanding the written business case and the spin-off benefits your project can generate for other projects and stakeholders, a PM can plan a project to maximize the delivery of business value beyond simple budget, time, and scope considerations. By getting in the head of the project stakeholders and factoring in political elements, too, however, the PM can move far beyond what is typically expected of the role, and he or she can gain the trust of the business and the career advancement that naturally results.

Kevin Aguanno

Kevin is a certified executive project manager with more than twenty years of experience helping manage large, complex projects in some of the world's largest organizations. He is a well-known expert on project management, teaching at several universities and speaking at conferences worldwide, and he has published more than thirty books, audiobooks, and DVDs on project management. He is a Fellow of the Project Management Association of Canada (IPMA Canada) and holds many project management qualifications. He is a pioneer in agile project management approaches and is heavily involved in the agile community. Subscribe to his free AgilePM Newsletter or get his agile project management tip of the week at **www.AgilePM.com.**

Kevin Aguanno, BA, MAPM, CSPM (IPMA-B), Cert. APM, PMP, PMI-ACP, CSM, CSP, FPMAC

Principal Consultant
GenXus Management Consulting

E-mail: kevin@AgilePM.com

Website: www.AgilePM.com

LinkedIn: www.linkedin.com/in/aguanno

Twitter: @KevinAguanno

Phone: 416-540-8570

CHAPTER 18

Project Failure Is Always Your Fault—or Is It?

By Harry Mingail

As a PMP and CBAP with thirty years of project experience, mostly turning around troubled projects, I'm here to tell you that it's not just you. It's the organizational "us." I've learned and succeeded in my career because there's a better way! I want to share with you my experience and tangible takeaways for your success.

Discover them in two distinctive yet frequently overlooked "secret" locations... your organization's Enterprise Environmental Factors and its Organizational Process Assets. They dramatically affect your project, positively as well as negatively.

Enterprise Environmental Factors (EEF)

EEFs are internal and external factors that surround your project. Just like cold, damp, and blustery weather environmental conditions, you need to preserve, protect, and adapt to your project's EEFs.

EEF Key #1—Organizational Culture

Accepted and shared organizational values, behaviors, attitudes, beliefs, and customs define company and departmental cultures. If one or more of your answers to the following questions is choice "a," your organization leans toward being an entrepreneurial culture. Under these circumstances, I have success being a project management minimalist... just key practices. Conversely, if you have more "b" answers, as I have experienced in large insurance companies, make contingencies for the reality of red tape in your bureaucratic culture.

1. Is your organization focused on a) results or b) rigid processes?

2. Does your organization a) empower decision-making or b) stifle personal expression?

3. Does your business a) encourage cost-justified continuous innovation and improvement or b) frown on change?

4. Are a) office politics and conflict or b) collaboration and cooperation the way people work?

EEF Key #2—Your Company's Strategic Plan

I was shocked to discover, as I was leading a corporate IT strategic planning project, that the organization had no written Business Strategic Plan. My encounter represents the nasty extreme of the challenges you have if your project attempts to deliver results in a loose, misunderstood, or nonexisting big-picture company strategic plan. Without a strategy, your project will be frustrated and buffeted by the interests of torturous decision-making from departmental silos less concerned about what's best for the entire corporation... the overarching strategy.

Problem-solve these symptoms, for example, as I did, by setting aside contingency time buffers. Document key elements of the undocumented strategy in the Project Charter as assumptions. And, if feasible, "put the monkey on executives' backs" by making the development of crucial strategic elements external dependencies to your critical path.

EEF Key #3—Regulatory Issues

Sarbanes-Oxley, ISO 9000, and Food and Drug Administration rulings are a few such examples. Regulatory bodies and their demands can radically affect your organization's business flows and, thus, your project. If you don't have good answers to these four key questions, I recommend that you get them so they won't ambush your project.

1. Do you know which business processes currently affected by regulatory challenges affect your project?

2. Is there a reliable audit trail of processes that have been changed—when and by which projects?

3. Does your project plan run contrary to regulatory or legal issues?

4. Are there pending potential rule interpretations that might affect your project?

While I managed a Basel 2 critical regulatory banking project, I unceasingly focused my stakeholders on tackling existing as well as anticipated regulatory interpretations. If I didn't, the bank's board of directors would have had to be fitted soon for striped suits and 8x10-foot prison cells.

EEF Key #4—Human Resources

Behind all challenges are people. People, as individuals and in groups, are always the key to victory. Human resources matter very much!

My projects walk the walk of my book The Extraordinary Power of Project Relationships, in which I underscore and expand the importance of interpersonal skills for any organization in Appendix G of PMBOK. The Appendix G list includes communication, cultural awareness, decision-making, influencing, leadership, motivation, negotiation, political awareness, and teambuilding. Managing your boss and discovering organizational staff stars and group creativity skills are among the other skills I encourage.

Reach for the low-hanging fruit first. Deal with what you can control—yourself. Be hard on yourself. Rate yourself in relation to the skills listed in Appendix G. Upgrade yourself where there are gaps.

As I did for a vital master scheduling manufacturing system project, strive to acquire them for the people you were allocated by the human resources department. If doable, get all these people in the same training session to be on the "same collective page." If you can't, play the role I tend to like—a "developmental manager" who nurtures by teaching, coaching, and mentoring.

EEF Key #5—External Industry Drivers

In each industry, manufacturing, banking, media, transportation, and mining are a few examples, your project operates within myriad external business drivers and conditions. Put on your "business hat." Competitive forces, trends, and currency fluctuations are some of the many to consider.

For instance, you'll steer your project better if you understand your industry's manifestation of the acronym Political Economic Social and Technological (P.E.S.T.) trends. Analyze corporate trends and their effect on you. Because trends, like risks, unfold in the future, I recommend you manage them as such. Apply avoid, mitigate, transfer, and accept responses to negative trends. Exploit, share, enhance, accept to the positive (yes, risks can be positive) ones.

- During a multinational insurance company project, my team addressed the perilous political ups and downs of the European Union.

- While managing a series of projects in a banking risk department, we dealt with economic trends of interest and inflation rates.

- During a project to merge three small media companies, we addressed social and demographic audience and readership trends.

- While managing a strategic product-sales scorecard project for a large fast-food multinational, we anticipated the evolving technological trends of hand-held devices used by distributors in trucks.

Organizational Process Assets (OPA)

Several prominent industry best practice areas, such as the following, emphasize the value of Organizational Process Assets:

- PMI's *A Guide to the Project Management Body of Knowledge (PMBOK Guide®)*

- International Institute of Business Analyst's (IIBA's) *A Guide to the Business Analysis Body of Knowledge (BABOK Guide®)*

- PMI's *The Standard for Program Management*

As per their definition of OPAs, PMI explicitly advises project managers to exploit them to "influence the project's success." Your organization maintains OPAs as policies, procedures, and guidelines that facilitate dependable performance. Project approval gates, performance review processes, and approaches to funding are some of the many OPAs I recommend you recognize and leverage to succeed. Overlook them at your peril.

Key OPA #1—OPM3®

When I take on a project, I immediately consider PMI's Organizational Project Management Maturity Model (OPM3®). Since OPM3® presents an organization-wide framework and rating system of portfolio management, program management, and project management, it helps me recognize the reality of what can enable or, conversely, devastate my project.

Lucky you, if your organization is at level five. Your organization has maximized the prospects for your success. At this level, measurable practices and sustained lessons learned continuously improve existing standards, methods, procedures, and staff. When you find, through your PMO, this to be the case, I recommend that you use it or lose it!

In contrast, at lower levels, particularly level one, practices constrict and confine rather than nourish and sustain your achievements. Level one tells me that I must be willing to be heroic like James Bond or Rambo to make things happen.

Don't just throw up your hands in despair. Fill the gaps with best practices and templates you know. Borrow from knowledgeable friends. Surf for templates on the Internet. When the going gets tough, good PMs step up to the plate and do the best they can with the reality they've inherited.

Need more information? Check out http://en.wikipedia.org/wiki/OPM3

Key OPA #2—Project Portfolio Management (PPM)

PMI's *The Standard for Portfolio Management* defines a portfolio as "The centralized management of one or more portfolios, which embraces identifying, prioritizing, authorizing, managing, and controlling projects, programs and other related work to achieve specific strategic business objectives."

As a project manager, here's why I care. Like you, my projects are buffeted by competition for funding, people, and material and equipment resources.

I know that PPM makes it more possible for executives to make timelier, fact-based, and rational decision-making for all projects in the best interests of the entire organization rather than silos. Frustrations and complaints increase, and morale dramatically diminishes, when your stakeholders understand that, for example, the project will be delayed because you lost valuable Jennifer and Jason for six months to a rationally determined higher priority project.

Unfortunately, PPM without supporting software is futile, which is why, when I learned that the government department that engaged me operated the PPM software package Clarity, I was relieved.

Your takeaway is that when your organization's priorities change, PPM is your best friend. When PPM doesn't exist, your organizational landscape hampers executive just-in-time, fact-based executive support relative to typically hundreds of competing priorities. You'll need to rely more on your sponsor to escalate effectively and power-sell issues resolution.

Key OPA #3—Program Management

When I was in charge of a corporate merger undertaking, I reached for PMI's *The Standard for Program Management*, not just the *PMBOK Guide*. You wouldn't use a hammer to cut wood! Don't use only the *PMBOK Guide* to try to succeed with what you think is a large project but really is a program. A program is a group of interrelated project and nonproject components better managed together to obtain benefits and controls that would not occur if they were managed individually.

My advice to you is to exploit a practical subset of the detailed practices associated with the following PMI program management themes, each of which go radically beyond their project counterparts: (1) Governance, (2) Stakeholder Management, and (3) Benefits Management.

Key OPA #4—COBIT

These days, nearly every project engages small, medium, and frequently large portions of IT. To the rescue comes Control Objectives for Information and Related Technologies (COBIT). Created by the prestigious Information Systems Audit and Control Association (ISACA), COBIT provides a maturity-level measurement rating and framework for IT, with emphasis on good IT governance. It strengthens IT alignments and execution toward realizing business, not just "techie" goals, objectives, realities, drivers, and demands.

Has anyone pointed the finger of blame only at the IT department lately? Here's your takeaway. If your CIO does not already have them, check out http://www.isaca.org/COBIT/Pages/default.aspx for assessment tools and methods to understand how your entire organization rates. You'll be rewarded with corporate-wide causes of IT project maladies, which in turn put you on the right problem-solving path.

Key OPA #5—Lifecycle Models

Engineering, IT, building construction, and other product life cycle models characterize stages, activities, roles, responsibilities, tools, and techniques.

Understanding and then ensuring that all stakeholders appreciate their life cycle role is one of the first things I do when turning around troubled projects. That's because, in the absence of an accepted life cycle methodology, expectations differ about work sequences—who does what, how, and when. Dysfunctional chaos, rather than harmonious and productive teamwork, jams the vacuum when life cycle methods go missing.

Next Steps

Beware the state of these key OPAs and EEFs. Help to change them if needed, and you can. Advancing by even baby steps enables your success. Sometimes, you can't change them in time for you to benefit. Under these circumstances, identify, problem-solve, and adapt to them. Treat each day as a blank page in your project's diary of accomplishments. Problem-solve external factors to turn your diary into your best possible project performance record.

Harry Mingail

Harry Mingail's (www.harrymingail.com), PMP, CBAP, and Six Sigma, books (Project Management Entrepreneuring, The Extraordinary Power of Project Relationships, Business Information Technology Strategic Planning), audios, and videos draw from his thirty years of freelance consulting and delivery of more than five hundred workshops and keynotes to business and nonbusiness as well as universities in North America, Europe, and Asia.

Harry Mingail, PMP, CBAP, Six Sigma

Business and Mathematics Degrees

Project Turnarounds, Workshops, and IT Strategic Planning

Website: www.harrymingail.com

CHAPTER 19

A Practical Way to Create and Sustain Client Ownership of the Project

By Robert K. Wysocki

Lack of client involvement has been identified as a major reason for project failure. I would add that it's not only important to have clients involved, but involvement must also be meaningful involvement. For the past twenty-plus years, I have used a simple homegrown practice in my consulting and training business that has meaningfully engaged clients, fostering an ownership position where clients do whatever they can to make the project successful! Remember, their reputation is on the line just as the project manager's is too.

In this chapter, I share my homegrown practice for creating and sustaining meaningful client involvement and present a real-life example of that practice. My hope is that you will integrate this with your practice and find yourself delivering better solutions with greater business value

My Model for Meaningful Client Involvement

Once upon a time early in my career as a project manager, I invited my client to work with me on a particularly complex software-development assignment my team was getting ready to start for them. The solution we looked for had been elusive for many years and had reached critical. We faced a particularly high-risk assignment because of the business process complexity.

This was the first time I had extended such an invitation, and I didn't know what response I would get. This all happened when software developers spoke to other software developers in acronyms that few outside their immediate circle understood. The business process we were trying to automate had never been automated. Now, the technology could support the business needs. It had always been done poorly, but now, there was a way to exploit the newest technologies and help the business process. This would be a new business system, and my team wasn't sure how to design the decision support system it required.

Back to the story. My invitation was an attempt to muster all the help I could. The client responded, "Oh, that project involves technical stuff. I'm not a technical person, nor are my people. Just let me know when you are done, and I'll look at what you have." That was a response from my distant past, and I didn't expect to hear it again. History did repeat itself, though. What a surprise!

We delivered a solution; it worked correctly, and it almost delivered the expected business value. You would be correct that we didn't deliver as good a solution as we could have. If I had known then what I know now after all these years, I would have refused to do the project without the client being meaningfully involved. That pushback was unheard of in those days. Client involvement amounted to nothing more than signing off on a successful acceptance test. Had I opted for that pushback, it would have been playing hardball and unacceptable to management. So, let me help you understand why I would do that today whenever those conditions arise again.

Project managers have learned (and the Standish Group has validated through countless surveys) that lack of client involvement is a major reason for project failure. I would add "meaningful" to the client involvement. I've known that for a long time, having discovered it myself as part of my learning experiences. It is even more significant now that we are firmly planted in the complex project environment where achievable goals and attainable solutions are problematic. In this brief article, I'd like to share what I have learned about attaining and sustaining client ownership and why I insist on meaningful client involvement in every project I manage.

The first and perhaps the most important advice I can offer you is to adopt a practice where you and the client co-manage the project. I have used this model in every project my company has ever engaged in with our clients. One manager is one of my consulting partners or I, and the other is from the client side. Both managers are equally involved and authorized to make all decisions and share in the success and failure that flow from their decisions.

Just think if you put your reputation on the line in a project. Wouldn't you participate in the project to protect your reputation and your business interests? You bet you would. I know I would, and I have in every project manager assignment over the twenty-year history of my consulting and training business. Besides delivering solid solutions, I have built partnerships with my clients. I am clearly part of their team!

So, the project is technical, the client is not, and they want to know why you want them as your co-manager. That's easy. Before the project was a technical project, it was a business project, and it needs a businessperson as a major partner and decision maker in the project. The project team should not be forced to make business decisions. As the technical project manager, you want every decision to be the best business decision possible, and your client is in the best position to assure that.

My client would hear me say that I wanted to do the best job I could, and it would not happen without their meaningful involvement as my co-manager on their project. In retrospect, my client co-manager participates in all decisions. They provide their product expertise while I provide the process expertise, and we do this as co-equals!

You need to keep the client in the best possible position to make those business decisions promptly. Given the need for a business decision, the project team can often present alternatives, maybe rank them, and even offer costs and benefits. Give the client whatever information you can to help them decide. Then, step back and let them decide based on whatever criteria they want to use.

In the complex project world, this is even more important and critical. In these projects, the goal, the solution, or both cannot be clearly defined at the beginning of the project. The search for an acceptable business outcome drives the project forward. Again, the client is in the best position to choose the alternative directions that lead to the deliverables that produce acceptable business value. The technical alternatives are presented to the client, and the client chooses the best alternative. These iterations are repeated until there is convergence on a goal and solution that achieve the expected business value, or the client ends the project because it isn't leading in a fruitful direction. The remaining time, money, and resources can be redirected to a more likely goal and solution.

This strategy speaks of a team/client partnership. Without it, success is unlikely. I have an example from a few years ago that illustrates my strategy perfectly. My client was an advertising executive for a prominent print weekly for the computer industry. His sales were slipping, and he needed something to draw companies to advertise in his classified section and readers to read his job postings.

A Real-Life Example

Here's another example from a real project I managed. It occurred during the Initiation Phase of a $5M, three-year project to build a web-based thin client decision support system. It was known that the solution would involve rather complex multidimensional algorithms whose business rules could not be defined clearly at the outset of the project, and so, the client asked to have prototype algorithms built. The algorithms would be tuned during the project so prototypes would be useful.

That responsibility was given to the Chief Developer. We'll call him George. George was a brilliant technologist with a keen business sense. His problem was that he was not a team player and impatient, and he usually added functions and features to the solution without discussing his ideas with anyone else, including me, the project manager. George reasoned that the client would obviously agree with what he had done. True, but his independent action and strong character prevented any other competing ideas from surfacing.

Initially, the client was somewhat reserved and uncomfortable working on a team of techies. The techies were uncomfortable having to dumb down their word choice. Everyone jockeyed for position in this strange setting. George's charm, brilliance, and sales abilities dominated the design process. We searched for a solution that had not yet been defined, and so, an open team environment with maximal client involvement was necessary if we were to explore every idea and encourage new ones. That could only happen if the client offered new ideas during the prototype process.

George was unaware that he was short-circuiting that process. I wasn't interested in the solution, according to George. I wanted the collective wisdom of the team and the client to discover the best solution. George's approach wasted time and caused more rework than would have been needed had he followed the agreed process.

Clearly, managing George was the key to get client involvement. I managed George with one simple practice. When problems or questions surfaced for team or client consideration, I had George agree not to be the first person to respond—to encourage the client a chance to offer their thoughts first. The client seemed to defer to George rather than contribute an idea George might shoot down.

George's delayed response eventually began to work, but I just had to keep reminding him of our agreement. It wasn't long before the client took over the prototyping sessions, and George made changes right in the prototyping session. The client became involved, and George took on the role of support to the client rather than leadership of the client. I could sense a more relaxed and open meeting as the client team members felt comfortable contributing without the feeling that George waited to pounce on what they said. The dynamics in the prototyping sessions were electric!

To the client's delight, the project finished successfully nine months ahead of schedule and $1.8M under budget. Readership of the Job Opportunity Bulletin (JOB) increased by 18,000 subscribers in the first six months, and with that came increased advertising revenues.

In Summary

The lessons I learned from this project were clear. No one can claim to have a corner on the knowledge market (that is, more than one SME is needed), and the client and every team member must be given a chance to contribute openly in a brainstorming fashion to the solution. Creativity is a critical component, and it must be openly encouraged and practiced. The technical team and the client team can form a formidable team, if given the chance, and exploit the resulting synergy. George's behavior impeded that formation; correcting it opened the door that allowed a successful outcome.

The project environment must be free and open. Ownership of the resulting solution can only come from giving all stakeholders an equal opportunity to participate meaningfully in the solution's development. I also learned that through ownership of the solution comes ownership of the implementation. Because it was their solution, they wouldn't let it fail. The client took the lead! How often can you claim that?

The practice I have shared with you is simple, but implementing it takes project manager leadership and courage. For some clients, that required selling the idea because they responded to my request saying they were not technical and couldn't contribute to a technical project. My selling proposition is that even though they might not be technical, I am not an expert in their line of business or business function.

So, by combining our separate expertise, we can produce an effective solution and create the expected business value that justified approving the project in the first place. They bring the business knowledge and experience to the table, and my team brings the technical knowledge and experience to the table. Together, we create the synergy needed to find creative solutions amid a complex project world.

Robert K. Wysocki

Bob has more than forty-five years' experience as a project management consultant and trainer, information systems manager, systems and management consultant, author, training developer, and provider. He has written twenty-two books on project management and information systems management. One of his books, *Effective Project Management*, 6th Edition, has been a bestseller, and the Project Management Institute recommends it for every project manager's library. The 7th edition is in preparation. He has made more than one hundred presentations at professional and trade conferences and meetings. He has developed more than twenty project management courses and trained more than ten-thousand senior project managers.

In 1990, he founded Enterprise Information Insights, Inc. (EII), a project management consulting and training practice specializing in project-management methodology design and integration, project-support office establishment, the development of training curriculum, and the development of a portfolio of assessment tools focused on organizations, project teams, and individuals. His clients include the largest corporations worldwide.

He earned a BA in Mathematics from the University of Dallas and an MS and PhD in Mathematical Statistics from Southern Methodist University.

Robert K. Wysocki, PhD

President

EII Publications
28 Vinton Street
Worcester, MA
USA, 01605

E-mail: rkw@eiicorp.com

Phone: 508-304-9695

The Keys to Our Success

Scope

The Keys to Our Success

CHAPTER 20

The Top Five Causes of Scope Creep... and What to Do About Them

By Elizabeth Larson and Richard Larson

It seems that the greatest challenges we faced as project managers related to scope creep. Projects we thought were small often grew and became difficult to manage. Our biggest lesson learned was that, to be successful, we had to manage scope, but to manage it effectively. To do this, scope management became more than simply dismissing a new feature or a change request because "it's out of scope." It became about determining the initial scope baseline and managing changes to that baseline.

Establishing the scope baseline was never easy. It involved the difficult tasks of eliciting the requirements from a variety of stakeholders with a variety of needs, understanding the requirements of the final product, getting consensus on which product deliverables to include in the project, decomposing those deliverables, and finally obtaining sponsor approval. Almost without exception, each of these steps proved more complex and time-consuming than originally planned, and each usually had to be repeated as new requirements surfaced. This chapter presents five common pitfalls that cause scope creep and shares lessons we have learned to help address difficulties in managing scope. We have included the perspectives of the sponsor, project manager, and the business analyst.

What is Scope Creep, Anyway?

When we began our careers in information technology in the 1980s, we didn't talk about scope or scope creep. There was no agreed use of either term. Today, at least, we have some common definitions for both. Here is the working definition of scope we will use in this chapter taken from *A Guide to the Project Management Body of Knowledge (PMBOK® Guide)*—Fourth Edition:[8]

- **Scope**—"the features and functions that characterize a product, services, or result produced in a project" (Introduction to Chapter 5). The project scope is the effort to deliver the product scope.

- **Scope Creep**—The addition of features or functions of a new product, requirements, or work not authorized after the agreed-upon scope. A common misconception is that expanding scope is scope creep. When projects expand, they do not necessarily suffer from scope creep. However, an unapproved expansion of scope does cause it, usually resulting in an increase in time and/or budget as well.

Why does scope creep occur? We asked our LinkedIn colleagues. Here are five answers we received:

1. Lack of clarity and depth to the original specification document.

2. Customers trying to get extra work "on the cheap."

3. Poorly defined initial requirements.

4. Beginning design and development before a thorough requirements analysis and cost-benefit analysis has been done.

5. Scope tends to "creep" when you do it to yourself because of lack of foresight and planning.

Change is not only inevitable, but also desirable. The trick is, of course, to manage it. Whether using an Agile or Scrum method or a Waterfall process, we need a consistent way to handle and approve changes. Remember that effective scope management means ensuring that all new features and functions requested are handled in a way deemed satisfactory by all stakeholders.

What Is Wrong with Scope Creep?

Scope creep is risky. Unauthorized changes are not usually analyzed sufficiently to determine how big an effort is needed to complete them or what the impact is to other areas of the project. And because we tend to be over optimistic (for example, "This is a small change. It shouldn't take too long") and eager to please, we often sneak these changes into the project, which means that stakeholders—the sponsor, testers, vendors, business and technical subject matter experts (SMEs), and others—are surprised by the unauthorized work. Surprises usually result in more work or additional costs and delays, and they are usually unwelcome.

How Does Scope Creep Occur?

Let's explore why scope creep occurs so we can discuss what you can do about it. Our Top 5 list of why scope creep occurs includes

1. Ambiguous or unrefined scope definition

2. Inconsistent processes for handling changes

3. Not enough time to manage requirements

4. Lack of sponsorship and stakeholder involvement

5. Project length

Ambiguous or Unrefined Scope Definition

Pitfalls

- **Sponsor View**: Projects start with a high-level vision that has been proposed in business terms. but not defined in project terms. Understandably, sponsors focus on their customers and competitors and do not think in terms of "scope" much less "scope management." Sponsors also do not think of their project scope as undefined. They have a vision and don't consider the myriad details needed to implement their vision. They have planned and thought about their business needs, sometimes for months or years; so, from their perspective, the project is well defined.

- **Project View**: Even when the high-level vision is documented in a Project Charter approved by the sponsor, the scope is insufficiently detailed until requirements are collected and a decomposed deliverable-based Work Breakdown Structure (WBS) is created. Without this detail, the team will usually build what it thinks is right with no or limited input from the business. Misinterpretation of the vision and expansion of scope usually results.

Lessons Learned

Clear, well-managed scope is a key to successful projects. Here is what we have found helps sponsors, project managers, and business analysts contribute to a clear scope:

Sponsors should create their business case and Project Charter, which can be a short document that contains the business need, the project vision, and high-level features in and out of scope. Project managers and business analysts might document business cases and charters, but to be most effective, the sponsor must articulate the business need and benefits contained in them. The business need from the Project Charter includes the business goals and objectives, the business problem to be solved, or opportunity to take advantage of. The business need describes the current situation, or "as-is." The vision and high-level features are the skeleton of the product, service, or result of the project, known as the solution, or "to-be." These are all items executives should be able to articulate on their own.

- Project managers should establish the project scope baseline by creating

    a. Project scope statements with both features in and out of scope

    b. A robust WBS that decomposes deliverables into work packages

253

- Business analysts can contribute to clear scope by

    a. Creating scope models, business process models, context diagrams, and use case diagrams, all of which help clarify the scope early in the project

    b. Documenting clear, complete, and concise requirements

The business need and desired high-level features from the Project Charter are used to flesh out the requirements of the desired product. Understanding which business processes will be affected by the project and which information will need to be added or changed, for example, helps clarify how big or small the project is. Having scope models are extremely useful in conversations with stakeholders about what is included in the project. For example, the project team might be unaware of how much a business area, such as shipping, might be affected by a new order-entry application. A scope diagram, such as a context diagram or process model, can help point this out.

Elizabeth worked on a distribution center project only to have Finance complain after the fact that the project changed some work products they receive from one system. Had we done a process model, even at a high level, we would have seen that a key output going from Receiving to Accounts Payable had changed.

Inconsistent Processes for Handling Changes

Pitfalls

- **Sponsor view:** Once project scope has been approved, sponsors view it like a contract for what will be delivered. Once set, few sponsors enjoy revisiting the question of what is in or out of scope.

- **Project view:** Most projects have many requirements that grow as the project progresses. In the day-to-day discovery and detailing of requirements, it is easy for stray and "rogue" requirements to be added, even without knowing it, resulting in these common problems:

  - **Unmanaged scope.** Without the control provided by scope management, project teams often decide whether to add or reject changes to product requirements. Scope then tends to be haphazardly handled, often without the knowledge of the project manager or sponsor.

  - **Unclear which requirements are in scope.** Often, business SMEs insist that requirements be added to the project. Usually working through the business analyst, these SMEs complain about how difficult the end-system or product will be to use if the changes are not included as part of the project.

  - **Gold-Plating** adds features perceived as useful or interesting beyond what was requested and approved. It occurs when someone on the team decides that a new feature would help the business without proper authorization.

○ **Scope "kill."** Project managers need to manage scope, not make scope decisions. However, when there is a rigid or cumbersome process for changing scope, project managers might inadvertently reject valid scope changes that would benefit the project. We call this "scope kill."

Lessons Learned

Project managers should create a scope management plan with a change management process outlining how changes will be authorized or rejected. This plan becomes part of the overall Project Management Plan approved by the sponsor. In discussions with the sponsor, the project manager needs to discuss the importance and contents of the scope management plan, so there are no surprises about the strategies that will be used to manage scope.

For example, if the plan says that every change needs to be linked to the business and project objectives in order to be included in the project, the project manager needs to review this strategy with the sponsor and get buy-in before putting it in place. This buy-in is easier when the project manager and sponsor work well together, when trust and credibility have been established, and when the project manager has the courage to provide well-analyzed advice and recommendations to the sponsor.

- Project managers should work with the business analyst to ensure that a requirements management plan is in place. This plan can be as formal or informal as appropriate to the project and should include a process to handle requirements changes. It is helpful for the project manager and business analyst to discuss areas of potential overlap between the two roles. For example, they need to discuss and agree who will write and who will approve the requirements management plan.

- To work effectively together, the project manager and business analyst need to collaborate. Each needs to understand that, for scope management to be effective, they need to communicate openly with each other. Business analysts need the courage to advise project managers about the scope as it relates to business impacts and dependencies. Project managers need to help business analysts understand the project constraints so the business analysts can help the business stakeholders understand those constraints to ensure realistic expectations.

- During project execution, project managers should follow the scope management processes outlined in the scope management plan.

Not Enough Time to Manage Requirements

Pitfalls

- **Sponsor view:** Sponsors focus on solving their business problems and, therefore, give little thought to requirements activities. They generally are not eager for the business analysts and business SMEs to spend time on these activities.

- **Project view:** The team often thinks that the deadline given them is unreasonable, so spending time managing requirements activities has lower priority than other project tasks. The team feels as if they're caught between competing interests. Project managers and business analysts understand the risks involved in not establishing a requirements baseline, documenting requirements, tracing requirements, or managing changes. On the other hand, they want to please sponsors who perceive that the team is unproductive if they spend time doing requirement activities instead of developing the product. Elizabeth remembers a sponsor telling her sternly, "There is no time for requirements. The project is late already."

Lessons Learned

- As the sponsor's trusted advisor, the project manager should clearly explain the importance of completing requirements activities to the sponsor. He or she should explain the risks of scope creep and the related cost overruns and time delays when requirements are unmanaged. Use examples from past, unsuccessful projects.

  As a project manager on a multiyear retail project, Elizabeth sensed the sponsor's frustration with the many requirements activities that spanned months. "I could ask the sponsor to blindly trust us," she said, but he was in no mood to do so. "It became necessary to point out the risks to the project and the organization of rushing through these activities. One large risk included the possibility that prices of items in all the stores could be wrong. The sponsor agreed that we needed to get the requirements right. When the project was in pilot, the same sponsor became delighted with the results, exclaiming that the requirements activities for the project took longer than any he had known. However, the development took less time than any project he knew of."

- The business analyst should:

  ° Recommend a requirements process that includes "just enough" rigor.

  ° Baseline requirements and then manage changes. Without a baseline, there is no way to determine whether a change is part of the original scope.

  ° Document requirements. Not having documentation leads to the amnesia syndrome (for example, "I don't remember telling you that...").

○ Trace requirements using a traceability matrix, a standard tool for ensuring that requirements align with the business need, goals, and objectives and, therefore, belong within the project scope. On a recent project, Richard found that different sales staff kept adding "indispensable" features to a new website. It became necessary to remind all the stakeholders of the project objectives and original scope. He explained how the original scope was set, how it amplified the business objectives, and how critical it was to implement the project sooner without these features than later with them. Requests for indispensable features all but disappeared.

Lack of Sponsorship and Stakeholder Involvement

Pitfalls

- **Sponsor view**: Sponsors are typically busy with many initiatives. They usually have too much to do and might not even know they are not "involved." Unless sponsors receive regular project updates, they tend to focus more on business, not project issues.

- **Project view**: Lack of sponsor and stakeholder involvement both have long been ranked the #1 and #2 reasons for project failure according to research (Standish Group, 2004, and again 2012). Part of the challenge relates to scope creep:

    ○ A disengaged sponsor is more likely to abdicate project decisions to the team. The less interested the sponsor, the more likely scope creep will occur because the nexus of activity moves away from the sponsor and toward the team.

° Stakeholders often don't devote enough time to project work. Their ongoing operational duties often prevent their involvement with project work. Nevertheless, their input is important to project success. If they don't participate, requirements are incomplete, or the project team fills the gap by guessing at the requirements. When teams fill voids created by underinvolvement, scope creep usually occurs.

Lessons Learned

- Sponsors should develop Project Charters to keep ownership where it belongs. The sponsor is in the best position to fully articulate the project vision, describe its benefits, and set boundaries. Sponsors should also request frequent status reports with just enough information, as they require.

- Project managers: Use project status reporting that engages sponsors, focusing on progress toward the deliverables. Quick, visual displays of status work well.

- Project managers and business analysts: Use a tool such as a RACI matrix (who is Responsible and Accountable, whom do we need to Consult, and whom do we need to Inform). This matrix helps identify stakeholders and manage roles and responsibilities on projects. Use it to clarify decision-making and to confirm SME participation.

- Project managers: Be sure to include lack of stakeholder participation as a risk, and build in contingencies for the likely occurrence of this risk.

Length of Project

Pitfalls

- **Sponsor view**: The longer a project runs, the more time sponsors have to refine their ideas, to compare notes with the competition, or simply for the business to change. All these are potential causes of scope creep. As sponsors ourselves, we can attest to having added unauthorized requirements to projects the longer projects continued.

- **Project view**: Our experience tells us that "The longer a project, the greater the chance of scope creep to occur." Large project scope and long project schedules open the door to additional product features and functions.

Lessons Learned

Rich remembers asking a vendor to complete a project in two pieces. The vendor decided against doing so. However, the vendor lost a key resource and reassigned others, extending the project far beyond the original deadline. Had Rich insisted on making that business decision, a critical piece of the project would have been implemented almost a year before it was.

- Project managers: Decompose projects into smaller subprojects. Agile methods such as Scrum deal with this issue by shortening delivery cycles and tightly controlling (time boxing) scope. Consider this method for appropriate projects.

- Business analysts: Break large requirements into small features. Have the sponsor and business SMEs prioritize these small features based on business value.

- Both: As milestones are completed, formally close them, conduct lessons-learned sessions, and celebrate, which helps maintain momentum and reinforces the results and benefits achieved. That in turn keeps sponsors and other stakeholders engaged.

Summary

Scope creep is dreaded on projects, and it happens for many reasons. Trying to prevent changes with an inflexible scope process will inevitably not only cause resentments, but also will result in a product or service stakeholders will not want to use. Elizabeth remembers telling a room full of stakeholders that they could not make changes because it was too near the deadline to do so. She learned that her role was to advise the stakeholders, not make decisions for them. Richard remembers having a scope baseline that was not detailed enough and which was open to various interpretations. Trying to figure out what the vendor had agreed to became contentious. So, we do need to manage scope, but we need to do so collaboratively, which means we need to be clear about an initial baseline, develop plans to manage changes, take the time to manage requirements, ensure that stakeholders, including sponsors, are engaged, and keep projects short so scope can be determined, agreed, and managed.

Table 20.1 summarizes our top five reasons for scope creep and a recap of what to do about each.

Scope Creep Cause	Recomm-endation	Sponsor Action	Project Manager Action	Business Analyst Action
Ambiguous definition of scope	Define and baseline scope	Create business case and Project Charter with specific product features	Baseline project by Create scope statement with features in and out Create a detailed WBS	Create scope models Define clear and complete requirements
Inconsistent scope management process	Manage changes as appropriate to the project	Approve changes and/ or approve authorizat-ion levels	Include a change management process in a scope management plan Ensure the creation of a requirements management plan	Create a requirements management plan with a process to handle requirements changes
No time for require-ments manage-ment	Explain importance to business	Negotiate a requirements management process that works for the business and the team	Ensure that requirements are collected and managed Negotiate a requirements management process that works for the business and the team	Recommend a requirements process Baseline requirements Manage changes Document requirements Trace requirements

Lack of business involvement	Establish ownership and authorization levels	Keep in mind that if you are not involved, the project will probably fail Take the time to write a Project Charter Request timely project status reports	Provide timely status reports Create a RACI to establish responsibility and accountability for project deliverables Add lack of involvement as a risk to the risk register	Develop a RACI related to the responsibility and accountability for requirements activities and business analysis deliverables
Project takes too long	Narrow scope for quicker delivery time	Prioritize subprojects and required functionality	Break project into smaller subprojects Recommend Agile methods as appropriate	Break large requirements into small features Help the SMEs prioritize the requirements

Table 20.1: Why Scope Creep Occurs and What We Can Do.

Elizabeth Larson, PMP, CBAP, CEO

With more than thirty years of project management, business analysis, and consulting experience each, Elizabeth and Richard Larson have presented workshops, seminars, and training classes on five continents. They have built and managed all aspects of Watermark Learning for more than twenty years, including the development of training curriculum, instruction, sales, marketing, and operations.

Their articles have appeared in *CIO* magazine, and they have been featured in *PM Network*. They were both lead authors of the *PMBOK Guide®* 4th edition, and Elizabeth was the content lead for the *PMBOK Guide®* 5th edition. Both Elizabeth and Richard were lead contributors to the *BABOK Guide®* v2.0, and Richard is a lead author on the *BABOK Guide®* v3.0. Elizabeth and Richard have coauthored *The Influencing Formula, Practitioner's Guide to Requirements Management,* and the *CBAP Certification Study Guide 2.0.*

Richard Larson, PMP, CBAP, President of Watermark Learning

Both Larsons are certified as Project Management Professionals (PMP) and as Certified Business Analysis Professionals (CBAP). Their company was awarded the PMI® Education Product of the Year award for 2011. The Larsons are proud grandparents of five lively grandsons. They love to travel, and they have visited more than thirty countries worldwide.

Elizabeth Larson, PMP, CBAP, CSM

CEO, Watermark Learning
7301 Ohms Lane, Suite 360
Minneapolis, MN
USA, 55439

E-mail: elizabeth.larson@watermarklearning.com

Website: www.WatermarkLearning.com

LinkedIn: Elizabeth Larson

Twitter: e_larson

Phone: Office +1-952-921-0900

Richard Larson, PMP, CBAP

President, Watermark Learning
7301 Ohms Lane, Suite 360
Minneapolis, MN
USA, 55439

E-mail: richard.larson@watermarklearning.com

Website: www.WatermarkLearning.com

LinkedIn: Rich Larson

Twitter: Rich_Larson

Phone: Office +1-952-921-0900

Quality

The Keys to Our Success

CHAPTER 21

Quality: They Can't Define It, but They Sure Know When They Don't Get It!

By Julie Grabb

Quality is the least understood project management topic, yet it causes project managers the most grief. How often have you and your team created exactly what was asked for, yet the primary stakeholders were still unhappy?

Quality is often thought of as subjective, in the eye of the beholder, something toward which to strive. How do we plan a project to achieve subjective results? We can't. Quality must be defined in specific and measureable terms. Easier said than done? It is as simple as specific questions to specific stakeholders.

My key learning

I have learned that the project manager must obtain specific details on what quality means to the decision makers. In addition, it is crucial to link the benefits to be realized to this definition of quality. These details evolve over time.

Once these details are known, quality can be built into the project plan and, by definition, increase the chance of success—success defined as the **clients/users/sponsor get what they want** out of the project.

Definition of Quality

A colleague convinced me to expand my idea of quality. We are used to thinking of quality as characteristics or requirements of tangible deliverables. How much does it weigh? What color is it? What functions can it perform? What is the error rate? He suggested that, although all these characteristics are certainly examples of quality criteria, I need to expand this to include project management deliverables, documentation, reports, and even processes. Don't these have criteria that must be met to make them acceptable or to obtain approval? Yes, they do.

Of all of the words used—specifications, functions, characteristics, features, expectations—one word can summarize them all—**requirements**. What do your stakeholders require of the project for it to be deemed a success and to be complete? Philip Crosby (1926–2001) gave as one of his four absolutes of quality: **Quality is conformance to requirements**.

Keeping this simple definition in mind helps minimize ambiguity around what the plan is meant to accomplish. Either we have quality, or we don't. Either we meet the requirements, or we don't. The hard part is identifying the requirements of the many stakeholders within our project.

- What conditions need to be in place before this project product is transferred into operations/maintenance/support?

- How will these criteria and conditions be met?

In *A Guide to the Project Management Body of Knowledge*© 4th Edition (*PMBOK Guide*©), this information is collected from the project statement of work (or other project triggering documentation) and translated into high-level information in the Project Charter. In Prince2©, this information is also collected from the triggering documentation or from a series of questions asked of the key stakeholders and is documented in the Project Product Description as part of the Project Brief. This Project Brief also contains the business case information.

What if the answers are unknown or incomplete?

I have learned that acknowledging the uncertainty/risk around the project end state greatly influences the approach taken to plan and execute the project.

Helping the sponsor and other decision makers acknowledge that planning is an iterative set of processes and conversations is the most difficult thing a project manager has to do. This idea is even more relevant when the final deliverable of the project is unclear.

Based on this uncertainty, the project might need to be approached as a series of iterative stages. Each stage fills in more information and detail. Before moving to the next stage, plans including estimates of time and budget are revisited and revalidated against a business case to ensure the project's continued viability. Agile project management methodologies as well as Prince2 embody this approach by way of the Manage by Stages principle.

It is also important to acknowledge that stakeholders have different perspectives on requirements. All are valid. However, not all can be met. There will be lengthy discussions and tradeoffs among requirements. When discussing the tradeoff among time, cost, and quality, the key question is:

- If we remove this requirement, how will it affect the ability of the final product to achieve the required benefits?

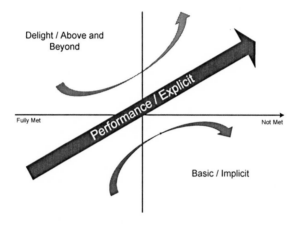

The Kano Model is a theory of product development and customer satisfaction developed in the eighties by Professor Noriaki Kano. This model divides attributes/requirements into three categories: (1) Basic/Implicit, (2) Performance/ Explicit, and (3) Delight/Above & Beyond.

Basic/Implicit Attributes

Basic attributes are the expected attributes or "musts" of a product, and they do not provide an opportunity for product differentiation. Very often, these attributes are unspoken, assumed, or "taken for granted." Increasing the performance of these attributes provides diminishing returns in customer satisfaction; however, the absence or poor performance of these attributes often leads to customer dissatisfaction.

The biggest difficulty with the delight factors is that what makes one person smile won't necessarily delight another person. In addition, something that created delight in the past might have turned into a basic or implicit requirement because stakeholders now expect it. You did it before so they assume you will do it again. If you don't do it or include the requirement again, they might be unsatisfied. Requirements that were previously delights might have turned into basic/implied, and your stakeholders have assumed they don't need to tell you.

- What would put a smile on your face if it were present?
- What was present in the past that delighted you and you now assume will be present again?

In the *PMBOK Guide*, this information is documented in the Work Breakdown Structure Dictionary. In Prince2, this information is documented in the Product Description for each product to be created. This includes project management products and those created by the specialists or subject matter experts.

Both include information such as:

- Description of work
- Assumptions and constraints
- Responsible organization
- Resources required
- Quality requirements
- Quality criteria
- Quality methods
- Technical references

Step 1: Begin with the end in mind.

Steven Covey said that highly effective people must begin with the end in mind. How often do you start a project without a clear vision of where you will end up? It is a *Catch 22* situation. Do we wait until we have a clear vision before starting any planning? If that were the case, not many projects would get under way.

I have learned that understanding the link between my project and the business situation leading to the need for this project helps me understand the big picture. The result is a clearer understanding among the decision makers about the results this project intends to deliver.

Best practice states that this information should include:

- The Business Need:
    - ° Why is the business undertaking this project?
    - ° What is the originating problem or business opportunity to be solved?
    - ° What business changes are expected when using the product/deliverables of this project?
    - ° What benefits are expected when using the product/deliverables of this project?
    - ° How does this project fit into the organization's strategic plan/goals?

- Project Product Description (final project deliverable):
    - ° What is the major deliverable (and component pieces) of this project?
    - ° What are the acceptance criteria for the project product?

But what if they don't know what they want? "I'll know when I see it," or "You're the expert; you tell me what I need."

As a project manager, you can help the stakeholders discover the right words to document their requirements. You can suggest wording, give examples, or ask pointed questions to get the details. You can combine your stakeholders' language with the specific detail needed for the project team. Remind your stakeholders that when they agree to the words you have given, they take ownership of the criteria.

Step 2: Identify quality for each project deliverable.

The project manager is not the only person responsible for identifying requirements. Often a business analyst is involved. For each deliverable in the WBS, quality criteria must be identified so we can establish

- What is included in each deliverable

- What the definition is of finished

- What the criteria are that will indicate this deliverable is finished

- How these criteria will be measured

- What specific skills are required to create and measure these criteria

As discussed earlier, not all requirements can be determined at the beginning of a project. As the project's scope is developed and further elaborated as the project progresses, these requirement conversations will have to be revisited.

- What do you assume we will include in the deliverable/project that you haven't told us about?
    ◦ Examples: documentation, adherence to appropriate legislation, translation, training, certain types of communications, warranty, testing, providing maintenance.
- What was missing on past projects even though you expected it?
- What will make you unhappy if it is not present?
- What are you taking for granted is included in this project?
    ◦ Example: brakes on a car.

Performance/Explicit Attributes

Explicit attributes are those for which more is generally better, and that will improve customer satisfaction. Conversely, an absent or weak performance attribute reduces customer satisfaction. Of the needs customers verbalize, most fall into the category of performance attributes. These attributes form the weighted needs against which product attributes are evaluated.

- What are your requirements?
- What do you expect to be included?

Delight/Above & Beyond Attributes

Delight attributes are often unspoken and unexpected by customers, but they can result in high customer satisfaction; however, their absence does not lead to dissatisfaction. Delight attributes often satisfy latent needs—real needs of which customers are currently unaware. Although they have followed the typical evolution to a performance and then a basic attribute, cupholders were initially delight attributes. Be careful that the delight factors do not cause the schedule and budget to be exceeded.

Step 3: Include planning details to ensure that quality will be delivered.

As planning and requirements discussions proceed, the quality details must be reflected in the relevant plans. Most important is to acknowledge all resources (both human and non-) required to create and measure the quality requirements.

I have learned not to treat quality, and all the work required to accomplish and check it, as an afterthought. Include it in the plan so time and money can be allocated to it and the stakeholders understand the cost of quality—the cost of doing it right—the first time.

Tasks included in the plan:

- Tasks to procure the appropriate raw materials and skills to create the required quality criteria.
- Tasks to check that the quality criteria have been met.
  - Examples: testing, walkthrough, visual inspection, checklists, sampling.
- Tasks to check that the acceptance criteria have been met.
  - These tasks will be scheduled near the end of the project right before handover of the final product is finished.
- Tasks for rework and rechecking.
  - If requirements have been unclear or uncertain, some rework will typically be required. Include this in the plan to start with as a risk mitigation action.

- Tasks for approval of deliverables.

  ○ This usually takes minimal effort as it will be based on the checking/testing results, but the duration must be accounted for, as the person with appropriate authority might need time to fit this into his or her schedule.

Resource Assignment on tasks:

- Include resources with the appropriate skills to create the required quality within the required timeframe.

- Acknowledge that the skill required to check something is not always the same skill needed to create something.

- Ensure resources who will approve deliverables have the appropriate authority.

Estimating:

- Include effort time (and therefore dollars) for the checking and rechecking.

- Include duration time for approvals—they do not happen instantaneously.

- Including dollars for the appropriate raw materials needed to build the required quality.

With this information included in the plan, quality is now one more thing on this list of things to accomplish rather than an afterthought.

Step 4: Include a discussion on the impact to quality when dealing with issues and change requests.

Remembering that everything is a requirement and, therefore, related to quality, when change requests are raised, the conversation must include any impacts to quality.

- *Example*: If the criterion for the time it takes to complete a new form is to be no more than five minutes, a request to add new items to the form should be analyzed to see how this would affect the time to complete the form.

- *Example:* If a skilled resource is going to be replaced with someone of a lower skill level, analyze the impact of the new skill level on the time required to create the required quality criteria/requirements.

- *Example*: If a request comes to cut back on testing to save time, analyze how not knowing fully if the requirements are met might affect the product in the hands of the project customer and, thus, the benefit realization.

Conclusion

As they say, the devil is in the details. So, spend some time getting them right. Do the right things right the first time. We always seem to have time to fix things later, but not the time to do it right in the first place. Quality might appear to cost more, but this is a short-term perspective. This short-term cost results in significant savings downstream and an increased chance of success.

Julie Grabb

Julie is an experienced project management trainer and coach with more than twenty-two years of progressive business experience in both the public and private sectors. Julie brings a proven success record in project delivery and training from both the public and private sectors. Bringing this experience into the classroom enriches the learning experience for all participants.

Julie is a veteran of more than 450 classroom delivers. Her participative management style provides effective leadership and fosters dedicated and competent project teams. Her enthusiasm and energy are infectious. Julie's strengths include facilitations, presentation, and training, as well as excellent written and oral communications. She is the author of *The Project Manager's Question Kit* series of books, and she has made presentations at project management events and PMI chapters around the globe.

Julie has a Bachelor of Mathematics degree from the University of Waterloo, and she has obtained a CGA professional designation. She is a Project Management Professional (PMP) and a Prince2 Registered Practitioner.

In March 2009, Julie became the first Canadian-authorized trainer for Prince2.

"I'm very excited to bring this great complement to the *PMBOK Guide* into Canada. Prince2 is a great addition to the framework detailed in the *PMBOK Guide* and contains many examples of best practices in project management."

"Every time I teach, I learn something new. Sharing and learning from the course participants enhances the learning experience for everyone involved. Continuous learning leads to continuous improvement for the participants, the instructors, and the curriculum development team."

Julie Grabb

J.A.G. Associates
Mississauga, Ontario
Canada

E-mail: julie@juliegrabb.com

Website: www.juliegrabb.com

LinkedIn: Julie Grabb

Phone: 905-302-6704

The Keys to Our Success

Risk

The Keys to Our Success

CHAPTER 22

Hope for the Best; Plan for the Worst

By Doug Boebinger

General Dwight D. Eisenhower, Supreme Allied Commander, World War II, was quoted as saying, "In preparing for battle, I have always found that plans are useless, but planning is indispensable." The reason plans are useless, as General Eisenhower indicated, is that no battle, no project, will ever run exactly according to plan—and your project will not be the first. Every project will vary and deviate from the plan, which should be expected. Actually, it should be planned, which is why the planning is indispensable.

One key lesson I have learned over more than twenty-five years of experience in running projects is that you have to commit to developing a holistic project plan, including proactively dealing with the project risks. In my opinion, project risk management sets apart the "generals" from the "privates" when being successful on projects.

Without project risk management, statistically, a project has a 25 percent chance of finishing successfully, defined as completing the agreed-to scope on time and on budget. We know this because, when using a standard bell distribution curve, there is a 50-50 chance of meeting the customer's schedule and a 50-50 chance of meeting the customer's funding requirements. Together, bringing a project in both on schedule and on budget is 50 percent x 50 percent, which equals only 25 percent.

I don't know about you, but I won't lay down a bet if I only have a 25 percent chance of winning. I don't know any customer, sponsor, or boss who would accept such odds! With a 75 percent chance of project failure, aggressive risk management becomes imperative if you have any chance of completing your project successfully.

This chapter will discuss a unique type of risks known as Black Swans. Black Swans are very, very low probability risks that have a very, very high impact. Black Swan risks cannot only cause complete failure of your project, but also for your organization and your personal life.

Black Swan events are game changers, not only for your project, but also for your organization and you. Because Black Swans strike without warning, traditional project risk management techniques, which rely on experiences to predict future events, cannot be used.

Black Swan Risks

There have been many Black Swan events over the past decades and centuries that have shaped the course of history and business. These events change the way people do business, view the world, and even run their personal lives.

To name just a few examples of Black Swan events:

- The Black Death in the late 1340s
- The invention of the automobile
- The Wright Brothers' first flight
- World Wars I & II
- The invention of the personal computer and then the Internet
- The terrorist attacks of Sept. 11, 2001
- The recent global financial meltdown

The name "Black Swan" has an interesting history. People in the Old World had always seen white swans. That was the only color of swan anyone had ever seen. There was no reason to believe that swans were any other color than white. There was 100 percent empirical evidence of this fact.

However, in 1697, while traveling around Australia, the explorers discovered the impossible—black swans. Their discovery destroyed the perception and everyone's belief that all swans were white. Since then, the term *Black Swan* has been used to describe something once thought impossible.

> *"But in all my experience, I have never been in any accident... of any sort worth speaking about. I have seen but one vessel in distress in all my years at sea. I never saw a wreck and never have been wrecked nor was I ever in any predicament that threatened to end in disaster of any sort."*
> —E. J. Smith, Captain, *RMS Titanic*[1]

---

[1] From 1907, just five years before the Titanic sank in 1912—the most famous shipwreck in history. You don't want to be remembered as the captain of your organization's *Titanic*!

Black Swan events were introduced by Nassim Nicholas Taleb in his 2004 book *Fooled by Randomness*. In his book, *The Black Swan*, he indicated that Black Swans exhibit three basic attributes:

1. **Outside the realm of regular expectations.** Nothing in the past can convincingly point to its possibility. No one believes it can happen because it has never happened. Black Swans are unpredictable. They are such surprises that they surprise even experts.

2. **Extreme impact**. The impact exhibited by a Black Swan is extensive. Global Black Swans, like those listed previously, affect just about everyone and every business on the planet. Organizational Black Swans can cause a devastating, if not fatal, blow to the organization. Personal Black Swans have an irrecoverable effect on you and your loved ones.

3. **Retrospective predictability.** People need explanations for why things happen. We don't accept "chance" as a reason. Human nature makes us want explain Black Swan occurrences—after the fact. The explanation makes the Black Swan explainable and predictable, retrospectively. Hindsight is 20/20, so we need to learn from past Black Swan events to develop responses for possible future ones (Black Swan responses discussed later in this chapter).[2]

No statistical distribution or predictive models describe Black Swans. The trick is you cannot predict Black Swans the same way you predict and plan for traditional project risks. Therefore, you cannot even roughly forecast the likelihood of their occurrence. Because they cannot be

---

[2] Taleb, Nassim Nicholas. *The Black Swan: The Impact of the Highly Improbable*. 1st ed. London: Penguin Ltd., 2007.

predicted, they cannot be controlled, which means you cannot prevent them.

In my career, I have had several situations on projects that could be classified as Black Swans. One such situation occurred when I had a project to install structural steel for a large industrial building. Without going into the details, a failure occurred on a column base while a steel worker was at the top of the column. He rode the column down to the ground, but sustained major injuries in the process. Several factors contributed to this failure—improper installation procedures, failed safety equipment, and an aggressive schedule. All these could have been prevented if proper risk analysis had been performed

The other was a project installing major steel beams, as tall as six feet, in an automotive stamping plant press pit. The beams had to be shipped from Europe through the Great Lakes and their navigation locks. The trick was that the project was performed in winter, and the Great Lakes navigation locks close in winter because of freezing. The contract could take time to be signed, which would dangerously encroach on the lock shutdown date.

Black Swans can occur in your personal life as well. Like most people these days, I know the sting of a sudden and unexpected layoff. Other personal Black Swan events include sudden illnesses or disability as well as early death. But even if Black Swans are unpredictable, it does not mean you cannot prepare for them.

Expect the Unexpected

To deal with Black Swans effectively, several proactive activities need to be performed at the beginning of the project and then reviewed throughout the project life cycle.

First, Black Swans have to be acknowledged as a possibility on your project. Ignorance might be bliss, but it ineffectively deals with Black Swans. Key project stakeholders need to know the possibility of Black Swans and be willing to not only discuss them, but also to put responses (discussed later) in place to deal with them.

In the case of the steel column failure, we had a first-aid station onsite and emergency numbers posted for police, fire, and ambulance. We had evacuation procedures in place in case of emergency, procedures to deal with the personal impact on the worker and his family as well as the business impact on the organization.

In the process of doing the traditional project risk management for your project, you need to create a special category for Black Swans on your risk register. During risk identification brain storming, these risks might be wild and very unrealistic, but that is exactly what should be placed in the Black Swan category.

Very high impact, worst-case scenarios need to be discussed openly and honestly. "It could never happen" should not be a statement uttered by anyone. (Remember the *Titanic!*) High-influence stakeholders need to know these scenarios and embrace their possibilities.

Of course, the question can be asked, "Can you develop a project plan with an infinite number of scenarios?" No. Of course, you can't. But that doesn't mean you shouldn't try to understand as many as possible.

One shortcoming of traditional risk management is looking at each risk as a standalone, individual event. Black Swans can develop from a "perfect storm" of familiar events occurring simultaneously. If you have ever seen the movie *The Perfect Storm*, you know what this means.

Over the past six years, I have traveled across North America doing various project management courses for a wide range of organizations. As I'm sure many of you can appreciate, nothing causes you to address risk management more than air travel. Flight delays, cancellations, crew scheduling issues and weather conditions can cause air travel to be a real headache.

One of the worst for me was trying to get from Detroit Metro to Halifax, Nova Scotia, Canada. It was supposed to be a simple layover at LaGuardia. It turned into a delayed takeoff from Detroit and missed connection in LaGuardia. Then, the rescheduled flight was cancelled. It was decided I would fly to Toronto to catch a flight to Halifax. An unexpected snowstorm in Toronto (that wasn't in the forecast) caused yet another delay.

Bottom line, I didn't get to the course until noon for the class supposed to start at 8:00 a.m. In conversations with the course participants, I had told them to have a plan in place for when I arrived for how we would get through two days of course materials in a day and a half. They did, it worked, and all left pleased with the training. Any of these events can be expected, but a combination of several almost sank the course.

Black Swan Responses

In real estate, the mantra is "Location, Location, Location." When it comes to Black Swans, the mantra is "Contingency Plans, Contingency Plans, Contingency Plans."

You might not know what could corrupt your data files, but that doesn't mean you shouldn't have contingency plans in place to deal with data corruption. You might not know what would delay a critical delivery, but that doesn't mean you shouldn't have contingency plans to deal with delivery delays. You might not know what could cause a key vendor to go bankrupt, but that doesn't mean you shouldn't have contingency plans to deal with the situation.

Concerning the press pit steel megabeams that had to go through the Great Lakes navigation locks, we had a Letter of Agreement signed to allow us to order the megabeams before the final contract being signed. Our shipping contingency was to take the beams through the Gulf of Mexico and up the Mississippi River and Ohio River and then truck to the stamping plant site. It would have taken much longer and would have been substantially more expensive, but shutting the project down for the winter and delaying the automotive start of production date was unacceptable.

Ironically, classic project risk-management concepts and techniques are not only beneficial for your projects, but also for your organization and your personal life.

When it comes to Black Swans, the best defense is a good offense. Here, the organization needs to be involved. The organization needs to promote excellence in its policies, procedures, and people. They need to fight inefficiencies continuously. Ironically, when business is going well, organizations and people become complacent. In this complacency, the environment is set for Black Swans. The organization needs to invest in capabilities to handle disasters that can affect the organization's most basic operations by establishing recovery capabilities, backup plans, and alternatives. Organization needs to "deposit" during the good times to "withdraw" during the bad times.

As for personal Black Swans, having the necessary insurance policies (life insurance, accident/disability insurance, health insurance, car and house insurance, and so on) is a financial necessity. Saving money for retirement, as well as potential no-income periods is essential. Keeping up on your professional skills, networking, and staying current concerning the developments in your industry are also necessary.

In other words:

Planning—Planning—Planning

Training—Training—Training

You need not look any further than military organizations to see the value of planning and training to meet very dynamic environments and mission statements.

And if a Black Swan were to occur on a project, you need to focus your team on the problem to drive to the solution. You need to implement your contingency plan(s) while keeping the team centered on the objective. This will help to reduce rumors and fears while working the solution to the Black Swan. Another aspect you need to handle while dealing with the Black Swan event is to provide sufficient information to nondecision-making executives. After all, you don't need any more "help" than necessary.

Black Swans are inevitable—are you ready?

Doug Boebinger, PMP

Doug has spent the last twenty-five-plus years helping many companies in various industries as a project manager, consultant, and trainer to implement project management and improve their project processes. He was the lead project-management consultant to Ford Motor Company's global Powertrain division for twelve years.

Doug received his undergraduate and graduate degrees from Purdue University in Construction Engineering & Management and Civil Engineering. He led the construction efforts on various exhibits at the ground-up construction of the Indianapolis Zoo, including the centerpiece Dolphin Pavilion.

Doug is Project Management Professional®-certified with the Project Management Institute since 1994.

In 1997, Doug founded Integrated Process Developers, Inc. (IPDI), an international project management consulting and training company providing project management training and consulting services to companies worldwide. Doug has created courses on a wide number of aspects of project management for a large number of companies in a variety of industries. IPDI is a Registered Education Provider® with the Project Management Institute® since 2001.

Doug is a past adjunct professor with various universities across the United States. He is an award-winning instructor for the highly successful Masters Certificate in Project Management program taught across Canada. At Cleary University, he was the Project Management Department Chair, and he created their original project management degree program.

Doug Boebinger, PMP

Integrated Process Developers, Inc. (IPDI)
1230 Sheridan St.
Plymouth, MI
USA, 48170

E-mail: dbb@processdevelopers.com

Website: http://processdevelopers.com

LinkedIn: www.linkedin.com/in/dougboebinger/

Twitter: @projmgmt

Phone: 734-207-5470

The Keys to Our Success

CHAPTER 23

Start a Project with Trouble in Mind

By Bob McGannon

One of the most important lessons I have learned in my career as a project manager is to assume that every project you are assigned to manage is (or soon will be) a troubled project.

Projects represent the manifestation of an idea that an organization's senior leaders might or might not universally share. I have experienced many cases where a project's stated scope and objective seemed straightforward and easy to understand, only to find senior leaders had differing views of the outcome and priorities for change. Thus, engaging in expanded communications and verification of stakeholder perceptions of the project—as you would do when diagnosing a troubled project—is prudent to ensure you get the project started on the right foot.

In addition to stakeholder perception differences, three other major characteristics indicate a "troubled" project. Cost problems, schedule problems, and scope problems where the project does not produce the product are the most common instances where a project will be placed in a troubled state. The causes for these problems are many and varied.

This is why, with all these potential challenges project managers face, it is wise to view each project with an objective and critical eye from the start of your project assignment. Treating every project from the start as if it's a troubled project will elevate your vantage point. From there, you can deal with any issues and the following potential danger areas early in the project lifecycle and settle them before they unsettle you and your project!

Potential Danger Area 1: Assessing Your Sponsorship

Sponsorship is an essential part of successful project delivery. Sponsorship can consist of a single empowered manager or a "steering committee" type of arrangement where several managers provide direction for the project. (In this chapter, we will use the term sponsor to refer to either of these scenarios.)

Managerial decisions are a regular part of moving through the project lifecycle, and they must be made efficiently to keep the project moving forward. All changes that result from a project's execution will have to be endorsed, with any new business processes supported appropriately. Hence, warning bell number one: without appropriate sponsorship, you are unlikely to get support from the business, and the benefits from the project will not be realized.

Consider the following sponsorship assessment statements that represent a situation where sponsorship aligns with the characteristics of successful projects. As a project manager, you should ask questions of the sponsor or your manager if any of these statements are untrue for your project.

Sponsorship Status Evaluation Statements

- The sponsor believes that the project is viable, should be pursued, and it is a priority in his, her, or their management responsibility.

- The project has sufficient priority in the sponsor's portfolio to warrant staffing, even considering other ongoing activities.

- The sponsor is willing to allocate the time to be engaged in the project as needed. If not, there are suitable delegates who can support the sponsor and the project manager to make business decisions and move the project forward.

- The sponsor has the ability to install and enforce the changes that will come about because of the project.

By considering these statements, you will quickly gauge if your sponsorship arrangement is not ideal. If that is the case, there are actions you can take to try to improve the situation.

- Talk to your manager, a key client stakeholder, or (if you can) directly to the sponsor and express your concerns. Hopefully, he or she can help influence the situation so a new sponsor can be found or a "sponsorship committee" can be created that will help drive the project forward. At times, more than one person needs to be involved to make the decisions necessary to progress the project. A committee is often a reasonable way to facilitate this process.

- In the instance where the current sponsor no longer has interest or priority to provide appropriate support for the project, you can partner with your manager to assess the interest of other primary stakeholders to take over the sponsorship role. If interest is not there, then the impetus to continue with the project no longer exists, and the project should be postponed or cancelled altogether. If interest is there, different sponsorship might mean the project's scope or breadth will change. As the project manager, you should be prepared for and participate in creating these scope changes to make the project redefinition and recovery process as smooth as possible.

- At times, the project sponsor has different priorities, but the viability of the project is still valid. In this instance, the project manager should work with her manager and the sponsor to determine if there is a way to move the project forward at a slower pace, or if there are project areas that can be progressed while others wait until the sponsor is ready to deal with them. If you go down this path, be sure to share the new direction openly and directly with the project's primary stakeholders. Raising any changes to the project's focus or timeline will help guarantee sound expectations management.

Potential Danger Area 2: Assessing Your Customer's Level of Interest and Understanding of the Project

The customer plays a vital role in your project's success. Through my career, I have identified a few customer-centric elements that should be evaluated to allow you to drive your project to a successful conclusion.

First, it's important to evaluate whether your customer understands the deliverables the project team intends to produce through the project and how they would be put into practice in their business area. It should be noted that:

- Interest and participation in a project can change because the business conditions that originally caused the project to be a priority have changed. A great lesson learned is to seek to understand if the business conditions that inspired the project have changed. If so, it is prudent to understand if the project, as currently defined, satisfies a business need that is a priority for your customer.

- A primary indicator of customer enthusiasm and buy-in is the degree to which the customer is willing to provide knowledgeable resources to help with the project. If the customer hesitates to provide those capable resources, a risk should be surfaced with your manager and potentially the sponsor, indicating a concern with successful project benefits realization.

Second, a critical lesson I have learned is that short-term customer behavior does not necessarily indicate a "permanent" problem for the project. Several alternatives should be investigated to determine if your customer's enthusiasm could be rejuvenated. My top three areas to investigate are:

1. **Timeframe:** Is the timeframe for receiving "value" an obstacle for your customer?

    - In many instances, there are specific benefits a customer hopes to realize, and the sooner they receive that benefit the more interested they become.

    - In many cases, a project can be separated into smaller phases, or specific components of a solution can be created. Determining these shorter-term deliverables and providing that business value earlier can be a way to invigorate the project and reengage the customer.

2. **Nature of the business**: Has the nature of the business changed in some way? Have the priorities within which the business operates changed significantly?

   - If substantial business changes have occurred, a new requirements session might be in order to realign the project with the business priorities it is intended to support.

   - If any new identified requirements are within the context of the original Project Charter, then a replanning of the project might be appropriate.

   - If the newly identified requirements are substantially different from what is described in the Project Charter, then the current project should be cancelled or suspended, and a new project should be proposed.

   - If a reason for ongoing viability for the project cannot be identified, the project should be cancelled.

   Note: Before undertaking any of these proposed actions, you should ensure that your sponsor is aware and supportive of a potential change in direction for the project!

3. **Customer participation:** Does the customer want you to move forward, yet they are not displaying the degree of participation you believe is appropriate for a successful outcome?

  - Working without an engaged customer adds risk to the project. It is up to you as the project manager to ensure that the customer is truly interested in the project and that they will be ready to implement or use the deliverables produced by the project.

  - If you determine that your customer is a "no-show," alert your manager and the sponsor to the situation and discuss the risks involved. If the sponsor agrees, proceed forward with the project, reviewing your directions and decisions with the customer and the sponsor frequently and in writing.

  - If the sponsor believes that project success is too much at risk without greater customer involvement, consult with the sponsor on how they can help reengage the customer or consider suggesting that the project be put on hold until this is achieved.

  - If the customer is not showing any signs of actively preparing for and setting up their organization to use the deliverables from your project, review this fact with the sponsor. My recommendation here is that the project should be put on hold until appropriate customer priority is reestablished or a qualified "change management specialist" is engaged to help you as the project manager and the customer organization to integrate your project deliverables into their processes.

Potential Danger Area 3: Assessing Your Project Team's Level of Support, and the Degree of Belief They Have Relative to Their Ability to Produce the Project Deliverables

An engaged project team that believes and understands the project's business value and what it will produce is critical for success in project delivery. It is well worth a project manager's time to work with project team leaders to ensure they support the project's direction and projected outcomes. Fundamental questions that a project manager can use to assess the team's support are:

- Does the team support the nature and composition of the deliverables they are asked to produce?

- Does the team understand the solution they are asked to produce and its relation to the business problem or opportunity addressed?

- Does the team have the abilities, staffing levels, and experience needed to produce the deliverables in the expected timeframe?

If these questions surface any concerns with the project delivery team, there are potential actions that can be taken to deal with the situation. These should be evaluated for applicability within your organization's culture, and you should include your manager in the assessment. You should also evaluate the extent to which the recommended actions are applied. Given those caveats, here are a set of proposed actions to consider for the most likely problem situations I have seen with project teams:

1. **Problem:** The project team is not supportive of the deliverables they are asked to produce.

   Solution: Determine the cause.

   - Is it because they believe the solution is high risk or overly complicated? If so, ask for a proposal of alternatives and convene a meeting with the team or person who designed the solution and surface the concern. A less complex solution could possibly be considered.

   - Is the dissention about the deliverables isolated to a few people? If so, try to determine what they fear about the solution or fear about their role on the project team. Working with your manager and the team members' manager to address those fears might help.

   - Is it because they believe other things are higher priority? Workload issues are quite common in many organizations, particularly with difficult to find skill groups. Seek to understand what the project team is working on. Determine if that priority seems appropriate, and review the situation with your manager and the sponsor. The sponsor can then choose to exercise his or her authority to rearrange priority or help alter the schedule for your project to align with the prioritization currently managed.

2. **Problem:** The project team does not understand the solution they are asked to produce.

   Solution: Convene a meeting with the designer of the solution and the team and ensure the questions the project team have are surfaced and addressed. A tie between the targeted business processes and the proposed solution should be clearly understood by all major project stakeholders.

   - If this cannot be resolved, all parties should review requirements for a common understanding. Often, misunderstood technical requirements are the cause of project issues, and they should be cleared up effectively with this review.

   - Is it because the team lacks the right team members or appropriately knowledgeable team members? Determine if you can get the right team members from within the organization or from contracted resources. If not, a risk exists, and it should be reviewed with your manager and the sponsor immediately for appropriate direction.

   - Is this new technology or a new approach? With your manager's help and support, work with the designers of the solution and have them work closely with—or as a dedicated part of—the project team. If this cannot be arranged, a major risk is present if the project moves forward. Discuss alternatives with your manager and the designer of the solution and present the risk and the potential solutions to the sponsor for a go or no-go decision.

3. **Problem:** The team does not believe they have the skills or experience to produce the solution.

   Solution: There are three possible ways you can solve this problem:

   - Determine if contracted skills are available to help.

   - Determine if the appropriate skills are elsewhere in your organization and what they are currently working on; reprioritization might be an alternative for the sponsor.

   - Determine if there are technical limitations in place that restrict the ability for the team to produce the solution. If that is the case, determine the viability of either incorporating the elimination of that restriction as part of project scope or suspending this project and launching another project to eliminate the restriction. If neither of these is feasible, then the current project might not be feasible and cancellation of the project is recommended.

It should be noted that it is **essential** to perform the evaluations recommended here in the sequence they are presented. The reality in every project is that a hierarchy of needs and authorities exists in the environment while you are trying to assess or recover a project. So, perform the evaluations, and consider the implications of these following factors that have the potential to influence your project's success:

- The sponsor's motivation and constraints for pursuing the project dictate the organization's limits in working on the project.

- The sponsor can "make or break" the actions needed to empower the project in terms of funding or prioritization of skills needed to make the project a success.

- The customer's desire to have the project completed depends on their being a sponsor with the appropriate authority to fund the project, designate critical team members, and prioritize organizational activities.

- In the instance where customer behavior needs to be altered, the project manager has little or no authority to instigate that behavioral change. Good leadership and influence practices (and your manager's help!) can, however, be applied. If those leadership and influence practices do not work, a sponsor is needed to help with this behavior change.

- Acceptance and participation from the customer is needed to ensure the project team can function efficiently and effectively without introducing undue risk to the project. The customer needs to be involved to ensure the solution produced by the project team is viable.

The sponsor's authority and constraints will affect the customer's prioritization and participation, which in turn affects the project team. Working with the project team or the customer without understanding the sponsor's constraints is a recipe for trouble; expectations can be set with the project team (or the customer) that the sponsor won't support. Thus, sponsorship must be assessed first, followed by the customer and then the project team.

Although assuming that your project is troubled from the start is not a panacea, in my experience, the evaluations outlined here will help you identify preventable project issues and empower you as you tackle projects with ever-increasing complexity. The lesson of starting a project and behaving as if it is "troubled" was an instrumental learning experience early in my career. Not only did working through and ensuring alignment among the sponsors, customers, and project team significantly increase the chance for project success, but it also enhanced how senior leadership and customers perceived me.

The confirming questions I asked to ensure alignment among and between stakeholders often surfaced solution considerations that otherwise could have been overlooked. Through these "troubled project" actions, I was recognized as having diligent thought processes and surfacing improvements to the business, even when I was only the catalyst for other's thoughts. I was perceived as a stronger leader and a more thorough project manager. This early lesson has served me well during my entire career.

Bob McGannon, PMP, MPC

Bob McGannon is vice president of Mindavation, Inc. in the United States and director of Mindavation Pty Ltd in Australia. Both of these consultancy organizations focus on increasing businesses' capabilities in the portfolio management, program and project management space by providing delivery consultants, basic through advanced training workshops, and coaching. Bob has worked with beginning to advanced program and project personnel with a wide variety of industry backgrounds from more than fifteen countries.

Bob specializes in strategic program and portfolio management, and he has helped set up project management offices on three different continents. He has twenty-five years of IT and project management and analysis experience, eighteen of those years with IBM Corporation.

His management experience includes technology outsourcing, IT delivery measurement and quality, business analysis, and general project management. Bob has worked internationally while playing a major role in the development of international trade applications in use in the United States and Europe. He recently worked in Australia on a consulting engagement with Medicare in Canberra. He has managed project teams as large as 460 people, once serving as the IT Delivery Manager for IBM, supporting American Express.

Bob is a certified Project Management Professional by PMI, is a certified Executive Project Manager with IBM, and a Certified Project Management Coach for the Boeing Corporation. He holds degrees in operations research and mathematics. Bob holds a practitioner's certificate in the PRINCE2 project methodology and an ITIL foundation certificate.

Bob McGannon

Director

E-mail: rmcgannon@mindavation.com

Twitter: @mindavation

Blog: www.mindavation.com/IDBlog

Phone: AUS 1-300-947-572
Phone: AUS 0-447-255-382 (mobile)
Phone: US 866-888-6463

Vision

The Keys to Our Success

CHAPTER 24

A Project Leader's Secret Tonic for Success— Project Vision

By Dale Christenson

Project Vision

If you don't know where you are going, you might just get there. One of the most important critical success factors to any complex project is the project manager's ability to create a realistic credible picture of the future so his or her project team can all try to arrive there. This picture is the project vision that represents the preferred future end state of your project and the change it will bring to the host or client organization.

In this chapter, you will learn the return on your efforts for taking the time to create a project vision. However, just creating a project vision is not enough as it is like looking, but not seeing. I have experienced projects that have flamed out or died for many reasons only to rise again from the ashes because of only a deep and shared belief in a project's vision. A project vision garners its power from being shared. The greater our ability to create widespread buy-in to our vision the greater the likelihood we will be successful in our project endeavors. In this chapter, we will briefly review four easy steps to creating this powerful project tool.

What Is a Project Vision, and Why Is It So Important?

In our research and practice, we have found that a project vision is a critical success factor for project outcomes. So, what is vision? The dictionary defines *vision* as "the ability to think about or plan the future with imagination or wisdom."[1] The etymology of vision derives from the Latin word *vid* "to see," the act of seeing, which is distinguishable from the Latin word *vis* "to look." Merely looking is quite different from seeing.

When **looking** at some projects, I have observed overtaxed project teams, but **did not see** that the burden was not purposeful goal-oriented work but frenetic make work that formed a façade of purposeful work. With the Latin root and the dictionary definition, we might suggest that one who envisions can visualize a future state.

There has been a plethora of articles, books, and audios on the concept of organizational vision and as

many definitions. Many describe vision as something that helps clarify the direction in which to proceed, suggesting it imbues a transformational quality that facilitates an enthusiastic and committed transformation from X to Y.

I have found it useful to define *vision* as simply a "statement of the preferred future state," but even this seemingly simple definition requires further examination,

---

[1] *The Oxford English Dictionary*, 2001, s.v. "Vision."

as we have two constructs in our definition: 1) the vision is future-oriented and 2) a vision is a statement.

This statement can be explicit or implicit. Despite their arena, vision statements can and are often entirely implicit and undocumented. The danger of the implicit vision statement is that it will not be understood or consistently conveyed to all stakeholders in an organization, which significantly limits the vision statement's power if it is not widely known and shared with all stakeholders. However, the idea of a preferred future state suggests that a transition is required from the current state to a preferred future state. The signaling of a change is one purpose of creating a vision, but there are many more.

A vision tells us what the results will look like and helps us in knowing when a project has finished. A review of project management literature supports the criticality of this aspect of project management. As Lewis explains, "If everyone does not agree on the vision, each person will try to achieve the outcome he or she imagines, (*often*)[2] with disastrous results."[3] Therefore, the creation of a common shared project vision is critical as a guide to future decision making, to manage scope creep, and to direct project activities toward benefit realization. Without this, a project is at risk of failure from the outset.

Common sense supports the wisdom in determining where we want to go. At first, a vision statement should inspire a cause or achievement of the dream, and it moves you through a process where you take ownership of the dream itself. Once project team members and all concerned feel a sense of ownership, they are likely, at their highest motivational level, to strive for project success.

As suggested, a vision must inspire stakeholders and project team members alike. A vision statement must

---

[2] Italics added
[3] Lewis, J. (2001). (p.117)

inspire passion. A passion that will see the project through when there seems at times little reason to go on. A passion that moves stakeholders and project teams beyond expected performance. A passion that transcends self-interest only to become self-interest. A vision sets committed direction and identifies a preferred future end-state. It provides a paradox of setting a temporary organization apart from the permanent (host) organization while unifying its workforce in a common purpose. A vision helps in future decision making, prioritization, and assignment of resources.

To bring about the required focus, a project vision needs; clarity, continuity, and consistency are arguably fundamental requirements of an effective project vision. An effective project vision statement should have the following key characteristics:

1. It must capture the core purpose, preferred future state, and essence of the project objectives, its *raison d'être*—it must be understood.

2. It must make a convincing case for following the project vision concept that can be internalized by project stakeholders and that provides a compelling value proposition—it must be motivational.

3. It must be consistent with stakeholder cultures or subcultures to appeal at the assumptions and values level so the vision statement artifact resonates with the stakeholder—it must be credible.

4. It should be proactive to facilitate teams to work smarter and more effectively, perhaps identifying stretch goals—it must be demanding and challenging.

A strong project vision was created for the British Columbia Government's Project Management Centre of Excellence that propelled this project to international success as shown by many awards:

All sectors of the BC Public Service employ effective project management practices that result in project success rates that exceed other governments as well as, industry standards, and provide the best value to the citizens of British Columbia. To accomplish this, we support our project personnel in their delivery of quality project management. We also seek to continually improve both organizational and individual project management professional practices to increase the overall maturity of project management within the BC Public Service.[3]

When considering the common-sense purpose of a vision, I have often found the acronym DRIVES™ helpful in logically grouping the varied purpose in creating a vision and its requisite attributes and characteristics.

Decision making—A vision needs to help contextualize future decisions and aid sense making. It helps strategic alignment, prioritization, and resource assignment.

Reason for being—A vision needs to identify the purpose of the organization and its uniqueness.

Integrate—A vision unifies disparate systems/ functions and unifies people toward a common goal with a common purpose.

Values—A vision needs to align with the core values of the organization as well indicate the value of the project. Typically, a vision statement will speak to the end benefit of the project for the host organization.

Empowers—A vision motivates and inspires people to achieve the purpose of the organization. It can free people to be creative and innovative within chaotic systems.

---

[3]Province Of British Columbia. Project Management Centre of Excellence

Strategic Direction—A vision identifies the strategic direction of the organization.

Now that we understand what the purpose of a project vision is and its necessary attributes and characteristics, we can look at a simple process to develop a project vision.

Developmental Process

Developing a project vision is a creative endeavor to paint a realistic, credible picture of the future that the project will engender. Building a shared vision is a process with four general steps that continuously improves on what it creates. The four broad steps are as follows:

1. Host Organizational Audit

2. Vision Development

3. Communicating the Vision

4. Continuously Improving

Host Organizational Audit

The first step is the Host Organizational Audit or termed by many an environmental scan of the current state. Here, we want to know where the vision will fit the host organization. What is the culture (values, beliefs, and assumptions) of the host organization? If the project vision does not align with the host organization, its likelihood of adoption is significantly reduced. And, at worst, it will be outright rejected for, as they say, culture trumps change, and any vision is a signal that change (desired future end-state) is desired and eminent.

The first step of project envisioning is a time of discovery whose data will transcend the envisioning phase and help with the entire project. Some questions a project manager would want to gain information on are as follows:

- Does the host organization have a vision statement?

- Does the organization have a mission or purpose statement?

- What are the host organization's strength and weaknesses?

- •What is the organization's primary culture (values, beliefs, and assumptions)?

- Is the expected change brought about by the project consistent with the culture of the organization?

- Does the host organization have any unique characteristics or competencies?

Information and answers to these questions help align the future vision statement with the host organization. It also lets the project manager tap in to various areas of special skill sets that might help in developing the project vision or the project itself. Last, this step is an anticipatory readiness stage "preparing for change."

Vision Development

The second step is where we **Develop the Vision** and vision statement. This can be accomplished using a linear or nonlinear approach. Linear approaches rely on logic and prioritization. This is typically completed through some form of SWOT or force field analysis with goal prioritization. The other option is to use nonlinear approaches that attempt to tap in to creativity. It is easy to see how the linear approach is called the scientific approach and the nonlinear, the art approach.

The best practice we have found is to create a project vision with a team of people, some whose orientation will be linear, whereas others will have an art orientation to bring creativity to bear on the envisioning process. To do this, we recommend that the team be asked the following types of questions before they attempt to create the vision or vision statement:

Linear

- What is the need?
- What is the solution?
- What is the gap?

Nonlinear

- What will our office/organization look like six months, one year, or further after the completion of this project?
- How will it be different?
    - ° What will be different within the organization once the project is complete?
    - ° What will be different for end users of this project when completed?
    - ° How will we know that this project has been highly successful?
    - ° In your wildest dreams, what would you like this project to achieve?
- Have we the right solution?
- If not, what might the right solution look like?
    - ° With the alternative solution, what will our office/organization look like six months, one year, or further after the completion of this project?

The power of a vision statement comes in part from its shared adoption by all stakeholders, and this adoption process begins by having these stakeholders participate in the development of the project vision. This is one time I suggest that a facilitator external to the project guide a group session in developing the vision. It is unrealistic to expect the project manager to be a participant as well as facilitate a group session of such importance. I also recommend that a vision session not be conducted without the project sponsor, as he or she is a key leadership figure in the project.

Implementation Phase

Once the vision statement is completed, reviewed, and approved, we are ready for the third step, which is the Implementation Phase. The development of a vision or vision statement does little to achieve project outcomes if the vision is not communicated, maintained, and continually improved. When considering groups of people or teams, Briner, Hastings, and Geddes say that the "most significant success factor for project teams is that they have a common and shared idea of what difference they are trying to make as a result of the project."[4]

In this step, we continue our analysis of who needs to be satisfied at the end of the project, and this reveals in part, to whom we need to communicate our vision. We also require the full support of senior management in adopting the vision. We must never lose sight that a vision garners its power from being widely shared and adopted, which also requires that the vision statement be widely communicated to all stakeholders consistently and repeatedly. This communication creates awareness of the vision and expected change.

We need to gain all stakeholders' buy-in to the vision. We achieve this when we see our vision in action, as action gives life to the vision. This begins again with our senior management. Stakeholders need to see that senior

---

[4] Briner, Hastings, and Geddes (1996) p. 89.

management is committed to the vision. They need to see senior management using the vision statement, referring to the vision statement, asking others about the vision statement, and making decisions based on the vision statement. Such action will begin to harvest widespread buy-in to the vision, and the power of the vision will be released with everyone in the organization working together in the same direction and same goal.

Continuous Improvement

The fourth and final step phase is the Continuous Improvement Phase. This phase should not be overlooked, and it is arrogant ever to assume we have created the perfect vision statement. Although a vision statement should remain relatively constant and unchanging, we must also be open to vision adjustments as the project progresses. An important part of the change or direction might have been left out of the vision statement, and this has alienated a group of our stakeholders. For this reason and others, we should be open to the idea of evaluation and reviewing our vision statement over time. If you determine there is a need to change the vision significantly, you might need to return to Step 1 and repeat the process of creating the vision.

As with creating a project, it is with creating its vision. You could certainly create a vision breakdown structure and become more detailed in how you create your project vision. However, the important points to remember are

- A project vision is a critical success factor to the project.

- The creation alone of project vision is good, but insufficient to promote project success.

- A shared vision is a powerful force to ensure everyone is pulling in the same direction with the same goal through to successful project completion.

In closing, the creation and propagation of a project vision statement requires a small investment of your scarce project resources but results in a significant and positive return on your investment. The development and implementation of the project vision is a best practice that DRIVES™ project success.

Dr. Dale Christenson, DPM, CMC, PMP

Dr. Christenson is the founder and president of the Project Management Centre of Excellence Inc. He is a Certified Management Consultant (CMC) and Project Management Professional (PMP) specializing in project management consulting and training. He combines an extensive academic background with real-world expertise in program and portfolio management, organizational development, change management, and critical project success factors. He's held project management leadership and executive positions in the public and private sectors, including executive director of BC Government's Project Management Centre of Excellence and assistant deputy minister of the BC Leadership Centre and Learning Services.

Dale has a Doctorate in Project Management from Royal MIT University in Australia and holds undergraduate and graduate degrees, diplomas, and certificates in criminology, counseling psychology, and project management. Dale has twelve peer-reviewed journal articles to his credit and two book chapters. He is a frequent seminar and keynote speaker at conferences. He is the winner of the Project

Management Institute's (PMI) Government Project of the Year Award, a finalist in the Canadian Institute of Public Administration Leadership Award, and winner of several teaching awards.

Dale is also involved in giving back to his profession. He has been the president of his local PMI chapter in Victoria British Columbia and Knowledge Lead for PMI's international Government Community of Practice, and he currently chairs his local CMC chapter.

Last, Dale strives to keep balance between his consulting and training practices, as he believes a balance supports and continuously improves each practice.

Dr. Dale Christenson, DPM, CMC, PMP

President
Project Management Centre of Excellence Inc.
Victoria, British Columbia
Canada

E-mail: dchristenson@pmcoe.ca

Website: www.pmcoe.ca

Phone: 250-217-5362

CHAPTER 25

Lesson Learned with Setting Project Priorities

By Michael Flint

Setting priorities should be one of the easiest things for a project manager. A defined priority is a key factor in a project's success.

We collect the requirements, interview the stakeholders, develop the project plan, and determine the schedule. With all the tools, the resources available, and the years of experience that preceded, the simple task of setting priorities should be a "no brainer." A simple question at the project's start of the project is what it really takes; ask, "Why are we doing this project?" The answer given and the person answering it will help provide the direction, scope, and focus required to set priorities and deliver success.

One of my clients wanted to launch a redesigned newspaper. The designers had done much work, many people had provided many ideas, and more were in development. When I asked the sponsor why we were doing the project, his answer set the scene. This national newspaper needed to be redesigned and updated with a current approach that added true value and be an alternative to the digital solutions available elsewhere.

Success for the project was success for the company. Failure was not an option. It was imperative that marketing, sales, editorial, and the other teams work together to deliver a superior product that would wow the current and future readership. The priority was clear. The scope and focus was set. The project had its direction.

As project managers, we are taught to follow a standard approach to initiating our projects—identify the sponsor, get the purpose and requirements, create a Project Charter followed by a project plan. Doing this is the standard approach, and it should lead to success. In my experience, if we follow the standard approach without question, we might fail.

My number one recommendation to anyone in project management is question everything! Do not just accept what you are told. Being told that the project is high priority without qualification is, in my mind, the same as a parent advising his or her teenager to do something "because I said so!" We all know that always leads to success—right? Getting a clear understanding why it is a priority makes all the difference. And if you do not know, ask!

The newspaper example above illustrates that knowing what the result was expected to be showed that the project's success was critical to the company's future. Not all projects will get as clear a priority setting as this, but the same approach applies. Ask why we are doing it. It could be this is the president's "pet" project. It could be a launch of a new product that if we meet the deadline, we will gain significant competitive advantage. It could be an infrastructure project that will be moving data, files, and users from one system to another—routine and normal (aka boring), but knowing that it is a 24/7 shop with no downtime allowed and a huge penalty payable if delivered late just changed the priority from normal to high, and high with significant challenges. Knowing makes a difference, so ask.

Not all projects are the number one priority, not all projects literally "bet the company" on the project's success, so how do we deal with this when the priority is not that clear or obvious? How do we avoid getting to the end of the project and finding that a task that should have been looked after has been missed? Avoid hearing our sponsor announce that we had missed a key deliverable? It does not happen all the time, but it usually happens.

And we all know that The Standish Group has been producing Chaos Reports for many years that show that projects fail more often than succeed. According to their studies, interestingly, the success or failure of a project is driven by many factors, the top three being user involvement, executive management focus, and clear statements of requirements. This suggests that these factors are those that really set the priorities for the project.

Projects by their nature are complex, multidisciplined, cross-functional, or they face many challenges, some predicted and expected; some not as easily identified. Priorities have a habit of going the route of the "squeaky wheel." In other words, if one group or department makes much noise about a particular feature or function, there is a higher likelihood that resources and focus will be given in that direction to solve their problem and just "keep them quiet," rather than an assessment that would examine the concern raised to determine whether it is a problem that can be addressed now or in the future, or if it truly is a high-priority issue.

Too often with the normal dynamics of a project, priorities tend to be lost in the deluge of issues raised and the number of risks. Sometimes, the routine of reviewing the Issues Log becomes the priority, rather than the review identifying which issue is the priority.

Sometimes, as experienced project managers, we miss the true status and the true priority. Our training calls for us to set priorities based on the triple constraints and which of the three is the most important to the stakeholders? Is it the **scope** of the work—the clear definition of what is delivered? The cost of the work—will funding determine the priority of the project? The **time**—will the world end if the deadline is missed? Add to this the other two constraints frequently skipped—*quality* and *risk*.

Quality can have a huge impact on priorities, as it directly provides the user experience for the product or service that otherwise was delivered on time and on budget. It is not unheard of that a project review shows that it delivered on time and on budget but failed to meet expectations. One reason is the project's quality control aspects. An example is the software company that drove to deliver their products based on key deadlines to beat the competition, although their applications would be full of bugs. The project manager's priority was the focus on the deadline. The company saw that as more important than quality, and they accepted the risk that delivery might result in complaints.

Risk is a priority for any project, but it is most often overlooked because we can always do a Risk Assessment and create a Risk Matrix showing how we will deal with that risk should the event occur. It tends to be routine, but if, for example, the company thinks that the deliverable is the last opportunity for the company to retain or gain market share, the project's completion is the highest priority. Sometimes, the sponsor might indicate that they are "betting the company" on the outcome. Knowing this is a huge indicator and potential motivator for the teams, suggesting that this is now really the number one priority for the company.

Which of the five constraints determine priority? The answer is any or all. A clear scope direction can set the priority just as easily as a statement about the risk aversion. An investment consultant will take a different approach with knowledge that the client is risk averse from the one who is more tolerant. Similarly, in projects where the sponsor is risk averse, the project must plan to address every potential problem as opposed to the project with a higher tolerance and their acceptance of risks.

Looking back at my projects and the lessons learned from them, among many observations, priority setting is a key influencer in achieving success. If the project is not a priority to the organization, it might not be a success.

Lesson learned from my experience in project management is the adage of keeping it simple. Despite whether we look at priority setting or another function of the project management spectrum, keeping the focus simple is a key approach. The metaphorical question asks, "How do we eat an elephant?" and the answer is, as we all know, "One bite at a time."

Many projects are "elephants," some with complexities beyond belief—compare a software rollout to launching the Rover on Mars. In both cases, the project manager in charge, with the help of the teams, set the priority for their launches, defined the work packages, defined priorities in those packages, and then started "to eat."

The best projects I have had were the ones where the priority was clearly defined. As an example, one three-year project that involved building a multifaceted training facility many years ago was successful based on our following two priorities set by the sponsor. These were:

1. To open on schedule and make it look as if we had been open for three months.

2. To be the most technology-advanced facility possible without making the place look like a technology showcase.

These priorities might be argued as visions, goals, or objectives, but regardless, these statements are very powerful. They were the guiding principles that allowed all plans and decisions made to be focused, which allowed priorities to be set within the project and outside the project concerning other competing projects.

If a task, subproject, or development did not deal with one of these two priorities, it was not in itself a priority task. The overall program and the subprojects were not simple, the complexity was all-consuming, but with these two priorities, the teams, the project leaders, the suppliers, and all concerned learned and knew that these two priorities would not be compromised. If a challenge was found, the approach was to determine whether it would affect one of the priorities, and if so, the focus was to determine how to overcome the challenge. If it did not affect, while important, it was recorded along with the many other problems and risks.

Another example concerned the particular and difficult launch of a very specialized product. The chief executive officer, as the real sponsor, made my life much simpler. He stated that we would launch on October 1, and my job was to make it happen! No what ifs, no discussions on budget, no resource assignment discussion—just meet the deadline, and get it done! Note that these omnibus statements carry an implied threat, but I have found that the expected impact of failure is assumed by real sponsors and the organization, which again allows the project manager to focus on the direction and not worry about losing his job.

One client provided his management team to help define the requirements. The project sponsor was the vice president for the division. The budget came from him, and he would make the financial decisions. In my review with the sponsor and hearing his commitment to remove any roadblocks for the project, just like any good sponsor, he revealed two key points. One was that the project would be expected to report to an executive steering committee

330

monthly (the project manager would be expected to provide that report to the committee). The second factor was the revelation that this was a "pet project" of the president. These statements changed the priority of the project significantly. The original project at the department level was all about the features and functions to be provided through the resultant project. Not a significant priority.

The sponsor revealed the potential visibility of this project, but it was unclear whether he saw it as a high priority and whether he felt it might fail and had the proverbial fall guy designated. The Steering Committee saw this as another project competing for resources, and unless they had a compelling reason and based on the project manager and not the sponsor reporting, this would not be a high priority. However, as this was the president's pet project, I managed to get an audience with him to get the direction and clarify the priority for the project.

Again, not all projects are high profile. Many do not have the "visibility" often used to describe a key project. So, how do we deal with a project that might not have the company's stamp of approval as a high-priority project? It is a priority for the project manager. The project needs to be delivered. Career decisions might be made based on the results. As project managers, we also need to be leaders. With no clear set priority coming from the sponsor, the organization, or the project description, you the project manager, as a leader, must provide the priority for yourself and your team. Without that, why would you or the team work on this project?

This calls for perhaps elements of "silver-lining" or "cup-is-half-full" approaches. There is a reason we do things; there is a reason we get out of bed in the morning and go to work. As a project manager, our role is to make a difference; we deliver something other people (excluding other project managers) cannot do; we know that *"If This Was Easy Someone Else Would Be Doing It"* ™ (MBF Consulting Services Inc.).

Based on these and many other truisms, we should be able to determine a reason we work on a project and set the priority in which we believe. Once that is known within yourself and your team, others will see that the team has determined a direction, a focus, and the priority. This is not a project that "has to be done," rather one that is "a challenging project that only we can deliver." The change in attitude affects the approach and the potential for success.

These priority-setting statements are incredibly powerful. They can provide the project manager in charge with powers of which we sometimes can only dream Referential power is the key feature, and working with what could be termed a critical or "pet" project for the organization or a top executive can be highly directional and a true motivator for the teams who will work long hours and will add more tasks to their daily lives than expected. The result is a team proud to say they were part of this project. Again, this is the ideal world in which a clear statement is given. In reality, this does not always happen, and it is found that stakeholders do not always agree with the project's priority.

For example, I was asked to run a simple audit of the readiness of an organization for change and provide an assessment on their tolerance to risk. In the six-week duration, I produced a report for my sponsor, the chief operating officer. This report highlighted many deficiencies and areas for improvement and some real issues that could cripple the organization.

It was passed to the president who essentially dismissed it. However, the topic was already listed on the agenda for the board meeting later that month. Enabling deniability, I am sure, I was asked to present my findings to the board. The chair stated that, although he did not like the message in the report, he understood what it implied and knew the organization had to deal with the issues, and quickly.

Given the go-ahead to deal with the problems, the chairman looked to the president and the COO and suggested that they make sure that I had everything available to me to deal with all the problems successfully. The priority of this project that had been a simple audit reporting exercise was now number one in the company, and it would be until my monthly reports to the board showed we had corrected all the problems.

Had this support not come from the chair, this project's future and its priority would be determined by the COO's belief in the required results. The project manager then could recognize and promote the project as the COO's number one project. Not a bad priority. At no point would I suggest to the team that the president did not support the project; the COO must deal with that challenge.

We have often heard that projects fail if there is no clear sponsor, and although this might be true, the project manager should be able to manage this situation by helping discover the project's priority relative to the organization, as was done in the above examples. To be successful, we need to find the highest point in the organization, find the person who has the stake in the outcome.

It should also be clear that, in reality, not all projects would be the highest priority, as with the above examples. They will have varying priority. The project manager needs to know that priority and must manage expectations based on it.

If the project is number 14 on the list, we can assume it will likely be late, might miss some features, or even go over budget. When faced with the prospect of a low-priority project, the project manager still needs to help the team deliver. Whether the project is listed as a particular challenge for the team to overcome or a key underpinning for the organization, the project manager can emphasize some factor to help the team focus. Getting resources, approvals, and so on for the organization will be a challenge, as the organization has deemed this a lower priority. By managing expectations and letting the decision makers know that the project will be late and miss deadlines if it is not fully supported, the acceptance of the delay can be used as a positive outcome.

For example, I had one client with a major infrastructure project we were delivering, but at every turn, the approvals were delayed or declined. It turned out that, despite the project's importance, the ongoing support of the production system was paramount, and the lack of approvals was based on only the potential impacts to production. Knowing that helped the team overcome the frustration, and we started to provide supportive statements when requesting approvals, showing how this either supported the production system or did not affect it.

Following the guides to project management, filling in forms to comply with PMO needs, or updating the latest report for Enterprise Project Management is not enough to determine the priority or success. We all need to be successful in what we commit to, and the ability to understand what that is leads to that success—whether it is a decomposition of the requirements, a detailed analysis of the systems, or the in-depth interview with the project sponsor. Knowledge helps. Knowing what it is, knowing what needs to be done, knowing who the beneficiary is are some key points.

The lesson to be learned is that if the project manager does not know what the project's real priority is, there is a high probability of failure. That failure could be avoided if the project manager spends more time talking with the sponsor and examining the requirements. No one wants to deliver a project to specifications and fail to deliver what was required.

Knowing makes a difference. Always ask why the project is run. Knowing the priority helps set and attain the scope, direction, and results. Knowledge is power; never accept the priority without question.

Michael B. Flint

Michael is a passionate, practical, resourceful project management expert, who has been recognized as an authority in project management circles. He frequently speaks on the subject at conferences worldwide.

He has an outstanding record of accomplishment for delivering results and benefits and meeting expectations for his many clients across most sectors of the marketplace. Michael qualified as a Project Management Professional (PMP) in 1998, and he has been a volunteer leader with the Project Management Institute since then. He is currently the senior vice president and past president of the Southern Ontario Chapter.

As a successful business-focused consultant and project management expert, Michael is the president and project consultant with MBF Consulting Services Inc. Michael's approach to his work, his projects, and his endeavors marries the practicality of the real world to academic theory. His view, attitude, and philosophy are captured in his branding tag statement: IF THIS WAS EASY, SOMEONE ELSE WOULD BE DOING IT.™

Michael B. Flint

MBF Consulting Services Inc.
6901 Barrisdale Drive
Mississauga, Ontario
Canada, L5N 2H5

E-mail: michael.flint@mbf.ca

Website: www.mbf.ca

LinkedIn: mbfconsultingservices
LinkedIn: Michael B Flint

Twitter: @mbf007ca

Facebook: mbf007ca

Phone: 416-931-6635

Did you like this book?

If you enjoyed this book, you will find more interesting books at

www.MMPubs.com

Please take the time to let us know how you liked this book. Even short reviews of 2-3 sentences can be helpful and may be used in our marketing materials. If you take the time to post a review for this book on Amazon.com, let us know when the review is posted and you will receive a free audiobook or ebook from our catalog. Simply email the link to the review once it is live on Amazon.com, with your name, and your mailing address—send the email to orders@mmpubs.com with the subject line "Book Review Posted on Amazon."

If you have questions about this book, our customer loyalty program, or our review rewards program, please contact us at info@mmpubs.com.

The Power of the Plan: Empowering the Leader in You

by Douglas Land & David Barrett

How many of us have offered to run a small project in our lives only to discover mid-way through that we really didn't have a good grasp of the art of managing a project?

How many of us would like to step-up and volunteer our time to run a project but feel we don't have the necessary tools.

This book presents a simple, easy-to-understand process for managing small, everyday projects. This is not "Project Management Essentials" or "PM 101". It is simpler than that.

We will take you through a series of ten steps for managing a small everyday project starting with the important question "Should I really do this?" all the way through to tips on closing it all down and celebrating at the end. We will show you how to run great meetings that people want to attend, how to create a simple schedule and budget, and more. Most importantly, we will give you the tools to deliver your project on time and on budget making you and others feel great about your results.

This book is here to empower you to say "Yes, I can run that event" and to give you the tools to make it happen.

Added Bonus: You will receive access to a website dedicated to sharing tools and templates contributed by us and our readers that you can use on any project.

Available from most local booksellers, Amazon.com or directly from the publisher at **www.mmpubs.com**.

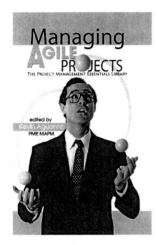

Managing Agile Projects

Edited by Kevin Aguanno

Are you being asked to manage a project with unclear requirements? High levels of change? A team using Extreme Programming or other Agile Methods?

If you are a project manager or team leader who is interested in learning the secrets of successfully controlling and delivering agile projects, then this book was written for you. From learning how agile projects are different from traditional projects, to detailed guidance on a number of agile management techniques and how to introduce them onto your own projects, this book contains the insider secrets from some of the industry experts - the visionaries who developed the agile methodologies in the first place.

Chapters focus on topics critical to the success of projects facing changing requirements and seemingly impossible deadlines. Chapters cover topics such as engineering unstable requirements, active stakeholder participation, conducting agile meetings, extreme testing, agile documentation, and how to use agile methods under fixed price contracts. The book also provides information to help you plan your agile projects better to avoid some common pitfalls introduced by the fast pace and concurrent activities common to agile development methods.

This book will show you the tricks to keeping agile projects under control.

Available from most local booksellers, Amazon.com or directly from the publisher at **www.mmpubs.com**.

Work Breakdown Structures for Large, Complex Projects
by Bill Bates

The organization of the plan for a large, complex project is critical to your success in monitoring, tracking and controlling that project. Do you know how to approach the organization of the work for such a project?

This recording will step you through such a process using a successful approach used by Bill Bates on a number of very large, complex projects. It is a simple, three-step Work Breakdown Structuring process that ensures maximum sub-tree independence. It also allows you to validate the correctness of the project structure as you proceed.

What Projects Should we Manage?
by Bill Bates

Every year, many major and minor project ideas are surfaced, discussed and either initiated or discarded. How are these decisions made? Is this a formal process in your organization that is connected to your corporate planning process, or not?

Project plans are operational plans that should be in support of corporate objectives and strategic directions. In order to select and manage projects properly, there should be a prioritization process. How does your organization handle that very tough issue? You need a project selection process that identifies, defines and determines the feasibility of each project and then passes the project through a selection screen that considers its support of corporate objectives, priority and benefits plus the availability of resources and funds.

Breaking the Rules of Project Management

by Paul Bergman

Trying to apply project management in a "paint by numbers" fashion does not work. With all the rules and guidelines of project management, one still cannot anticipate project success if only everything were done according to the rule book. Sometimes, the most successful project is a result of knowing when project management rules should be followed, and when they should be broken.

This recording will explore some of the more common project management rules and guidelines, and it will discuss what conditions might warrant a variation from these rules. Specific topics covered include activity durations, work breakdown structures, communications, and stakeholders.

Available from most local booksellers, Amazon.com or directly from the publisher at **www.mmpubs.com**.

Telling Truth to Power
by Catherine Daw

The failure of projects remains a critical challenge. Reports such as those done by the Standish Group or Gartner indicate that more projects are either canceled or fail than are actually ever delivered successfully. In fact, the failure to success ratio is almost 2:1. And, this doesn't even speak to the category of challenged projects.

As we delve into this issue we touch upon a key success factor: top management support. Looking at this from two dimensions we understand that for projects to be successful it requires open and honest communications with upper management. This in turn requires Project Managers to be able to deliver status – good or bad – without fear... herein lies the dilemma.

This recording discusses typical barriers and possible solutions. In addition, it provides key insights into gaining top management support through better communication.

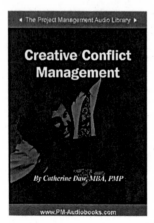

Creative Conflict Management
by Catherine Daw

Conflict is inevitable on any project but is it unpredictable? How you anticipate and deal with it will make the difference in terms of project, product and team success.

This recording focuses on creatively managing conflict in today's complex and stressful project environment. You will learn the different conflict management styles, self assess your style and how to adapt depending on the situation, understand the various sources of conflict during the life of a project, and develop proactive tactics to manage the inevitable conflict that will occur during a project.

Your Future as a Project Manager

by Frank P. Saladis

A question that is continuously being asked by project managers in every industry is "What demands will executive management expect from me and what skills will I need to maintain my value within the organization?" This is a question that addresses the future state of the project management profession and emphasizes the need to consider the rapid changes we are all experiencing. The future is, in most cases, unpredictable but the project manager can prepare and plan for changes as the business environment continues to evolve. This planning includes the identification of key competencies and a process for continued professional development

This recording focuses on the current trends impacting the professional project manager and provides some key strategies and steps that a project manager can take to achieve success as World Class Project Manager.

Available from most local booksellers, Amazon.com or directly from the publisher at **www.mmpubs.com.**

Effective Virtual Meetings: Tips for Teleconference and Webinar Leaders

by Claire Sookman

Virtual team meetings (teleconferences, webinars, video conferences, etc.) are increasing in popularity as organizations look for ways to decrease costs and increase productivity, service delivery, and employee well being. Virtual meetings can be as effective as face-to-face meetings, at only a fraction of the cost; however, the effectiveness of these meetings is constantly threatened – 70% of participants in teleconferences and other virtual team meetings admit to being distracted during the events by other work activities.

This audio recording of an interview with Claire Sookman, a leading expert for over 15 years on virtual team management, reveals solutions to many common virtual team problems. She shares:

- How to use language effectively to facilitate communication and participation

- 6 ways for reducing distractions during virtual meetings, keeping participants engaged and increasing the flow of information

- How to create more structure for virtual meetings to facilitate team bonding and collaboration

- 3 techniques for dealing with challenging or difficult team members

- 5 additional ways to encourage collaboration and build relationships

- 3 things to avoid when managing virtual teams

Available from most local booksellers, Amazon.com or directly from the publisher at **www.mmpubs.com.**

ACROSS THE HALL
AROUND THE WORLD

CLAIRE SOOKMAN AND SUSAN GARMS

Across the Hall, Around the World: Teambuilding Tips for Distributed Businesses

by Claire Sookman & Susan Garms

Having trouble engaging and energizing teams across your business? Many people are struggling with motivating teams, especially ones where some team members work in different locations— even across the world.

In this book, Claire Sookman and Susan Garms, experts on motivating distributed (a.k.a. "virtual") teams, bring you a number of teambuilding activities and exercises that you can put to immediate use.

The techniques and exercises in this book will help you

- Build collaboration
- Help team members get to know each other
- Conduct more effective meetings
- Bridge the gap between different cultures, and
- Overcome the obstacles of working in virtual teams

Available from most local booksellers, Amazon.com or directly from the publisher at **www.mmpubs.com.**

Project: Impossible — How the Great Leaders of History Identified, Solved, and Accomplished the Seemingly Impossible — and How You Can Too!

by Michael Dobson

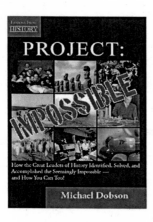

Nothing is impossible if you have unlimited time, resources, and flexible objectives. Project managers never find themselves in such a situation. Our projects are impossible if they can't be done within the constraints...but sometimes there's a way around even the most challenging barrier.

What can you do when the situation looks hopeless? In this exciting journey through history, you'll learn how the greatest leaders and project managers of the past took on impossible challenges...and succeeded.

- At the Battle of Alesia, Julius Caesar was outnumbered, cut off, and completely surrounded. His strategy? Change the game from fighting to digging.

- When he wanted to be first to fly nonstop to Paris, Charles Lindbergh was up against competitors with more funding, more experience, and over a year's head start. His strategy? Rethink the thresholds of risk.

- The other generals laughed at George Patton when he offered to send two divisions to rescue the Battle of the Bulge in only 48 hours. His strategy? See the future and get ready for it early.

- For mission director Gene Kranz, the odds against a successful rescue of Apollo 13 were daunting at best. His strategy? The Kranz Dictum, a powerful strategy to deal with crises even before they occur.

In PROJECT: IMPOSSIBLE, you'll learn a step-by-step methodology to succeed when facing even the most difficult projects. You'll learn Dobson's Laws of Project Management and discover the Godzilla Principle. From redefining the problem to challenging the project parameters, you'll know how to attack a seemingly impossible project...and get the job done.

Top-Gun Project Managers: 8 Strategies for Reaching the Top of the PM Profession

by Richard Morreale

Ever wonder why some people's careers rocket upwards while yours feels stuck, with you repeating the same drudgery day after day? If you are a project manager (or are interested in becoming one) then this book was written just for you. It shares eight strategies that you can use when plotting your career path—the trajectory of your own rocket—to help you reach the stratospheric levels of the profession.

These eight strategies are not just high level theory. The author of this book, Richard Morreale, has put them into practice successfully in his own career. Morreale is one of the top project managers in the world, specializing in turning around some of the nastiest, largest, troubled projects you will ever find; in fact, Morreale is often called "the Red Adair of project management" after the famous engineer who specialized in putting out oil well fires. Morreale's career has spanned a wide range of projects from working as part of the Apollo Program Team, helping to put men on the moon (and getting them back), to working as part of the team that computerized the UK Income Tax System. He also led the rescue and delivery of a $450M program for the 43 Police Forces in England and Wales and directed programs of work for some of the largest companies in the world.

Read this book to learn how you too can copy Morreale's career success—with these eight strategies, the sky's the limit!

Available from most local booksellers, Amazon.com or directly from the publisher at **www.mmpubs.com.**

CPSIA information can be obtained
at www.ICGtesting.com
Printed in the USA
FFOW05n2001200913
1855FF